WEST SHROPSHIRE
MINING FIELDS

Some members of the Shropshire Caving and Mining Club 1981, organisers of the first national mining history conference held in Shropshire in association with the Ironbridge Gorge Museum. From left to right: R. Taylor, D.R. Adams, J. Heathcote, A. Taylor, G. Trystman, R. Savage, M. Moore, R. Anderson, I.J. Brown (author), C. Trystman and M. Clough. (H.M.Parker).

The Shropshire Caving and Mining Club

While a group known as the Shropshire Mining Club and a number of individuals had been researching the history of the county's mining areas in the 1950s, it was not until 1961 that the present group was formally established. Some forty years have now passed, nearly 300 newsletters, journals and special publications have been produced, most mines have been visited underground and on surface, in some cases hundreds of times but still the sites hold a considerable fascination – and fresh discoveries of artifacts and tunnel extensions are being made all the time. The Club's strength has increased over the years, with at present over 150 individuals 'paid up' and very few weeks pass without at least one organised activity.

The Club is a strong supporter of the National Association of Mining History Organisations and conforms with all the established codes of practice for health, safety and the environment. The addresses of its 'officers' change frequently but the present contact address can be obtained through NAMHO's mailing base at the Peak District Mining Museum, Matlock Bath, DE4 3PS, telephone number: 01629 583834.

Note: Old mines are potentially dangerous and should not be explored underground without an experienced guide. Permission should always be obtained before venturing onto private property and the mention of a particular site does not imply any right of access.

WEST SHROPSHIRE
MINING FIELDS

Ivor J. Brown

To Kenneth Lock and the late Malcolm Newton,
and all members, past and present,
of the Shropshire Caving and Mining Club.

TEMPUS

First published 2001
Copyright © Ivor Brown, 2001

Tempus Publishing Limited
The Mill, Brimscombe Port,
Stroud, Gloucestershire, GL5 2QG

ISBN 0 7524 2363 0

Typesetting and origination by
Tempus Publishing Limited
Printed in Great Britain by
Midway Colour Print, Wiltshire

A kibble of lead ore at Perkins Level, Snailbeach 1965. This was probably some of the last ore produced, obtained as a by-product of barytes working during the mid 1950s. A wooden wheelbarrow lies against the shed of timber and corrugated sheet. Large metal kibbles, or hoppits, were still being made for sinking shafts and raising ore by Shropshire's Lilleshall Co. in the 1960s.

Contents

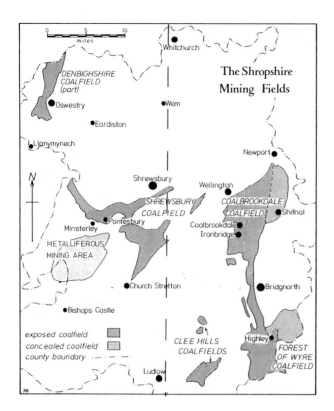

The Shropshire Mining Fields. The area covered by this book is shown on the left hand side of this map.

Acknowledgements

This collection of photographs and images is the result of nearly fifty years work; over this time so many people have been involved it would be impossible to name them all. But to everyone with whom I have been in contact, as local residents, historians or officers of organizations or groups, I give my thanks.

In particular I would like to thank Ken Lock for reading and commenting on much of the text and for making his large collection of slides and prints available. Also to Mrs Sheila Newton for permission to use Malcolm's drawings. For many years our specific interests complemented each other. Ken would collect anecdotes and old photographs, Malcolm would draw and I would research and write. I am especially grateful to them as, for the purposes of this work, they were able to step in and provide replacements when many of the my own irreplaceable 1960s slides were lost in a mugging in Spain in 1996. Others who have contributed considerably to this work include Andy Cuckson, Kelvin Lake, Jane Price (librarian to the Shropshire Mines Preservation Trust), the 'Never on Sunday' Project Team, Shropshire Records and Research Service, Terry Davies, Mrs Marcia Fletcher, Mrs Elsie Rowson and Andy and Kim Yapp.

Thanks are due to all who have provided photographs and illustrations, acknowledgements for which are made beneath their respective contributions. Photographs not credited are from the author's collection.

A special word of appreciation must be given to my wife, Iris, who has typed, checked and given encouragement throughout the preparation of this and many other 'mining publications'.

Introduction

It is not only for the convenience of this work that Shropshire has been cut by a north/south divide – for both east and west have their own industrial and mining histories. East Shropshire is already covered in a sister publication; it is an area of a few highly concentrated and, in their time, intensively worked coalfield areas. West Shropshire, on the other hand, has remained mainly agricultural even in its mining areas which are generally small and hilly. One area, very hilly and mineralised, turned out to be rich in metal ores and fortuitously placed alongside the Shrewsbury Coalfields, the other small mineralised areas never produced much ore. In the northeast, around Oswestry, the Denbighshire Coalfield extends into the county, this area has always been more closely associated with North Wales than Shropshire. It is intended that this volume will cover in detail the south-western area, the orefield around Shelve and its neighbouring coalfield near Shrewsbury, but examples are given from the northeast for completeness.

For its area Shropshire has been both rich in variety of minerals available and also in its geology; eleven of the thirteen recognised periods of geological time are to be found here. Many eminent geologists have worked in the county, often attracted by its mineral industry, and one of the foremost, R.I. Murchison, who did much of the early work on the Silurian System, spent so much time in the Hope Valley that a mine access road was named after him in his presence.

In Eastern Shropshire, the sedimentary rocks, the clays, ironstones, coal and limestone were the principal products of importance, but in Western Shropshire it was the 'metalliferous' minerals, the ores of lead and zinc, barytes, calcite with small amounts of copper, witherite, fluorspar and even silver, that were significant. There was also some coal, conveniently situated near the metal mining area, as well as further north.

The western part of the county has also been endowed with quarried products, sandstones, igneous rock and limestones and, equally importantly, sand and gravel. Peat has also been worked in northeast Shropshire and some exploration for oil and beds of salt have recently taken place.

Although the metalliferous ore producing area has been small its production has been significant both nationally and to the local economy. The lead mining industry, for example, seems to have been of some interest to the Romans and, in more recent times, Shropshire was a major producer of barytes, at one time producing up to a quarter of Britain's supply.

The employment provided by the working of the minerals has been important, although never large, but useful bearing in mind the lack of alternative work. The metalliferous mining industries probably never employed as many as 1,000 at any one time, although it must have neared such a total durng the peak years of the 1870s. By 1890 employment levels had fallen to less than 400 but rose slightly to about 500 in 1900. From this time they were in terminal decline, disappearing entirely in the 1940s and 1950s. Normally there were no more than twenty mines at work at a time and even at peak times only five or six mines employed over fifty workers. In the 1870s there were only eight mines employing over ten men and just three employing over 100 (Snailbeach, Roman Gravels and Tankerville).

The coalfield area around Shrewsbury has been even smaller in terms of employment and productivity – no more than 350 men, at up to ten quite small mines, seem to have been employed at any one time, although by the 1930s amalgamations had left just one mine employing about 200 men. A similar situation occurred in the Oswestry area coalfield – but the last mine survived to 1968 and over 1,100 were employed, the largest mine Shropshire had ever known.

A problem in the search for old images of the area has been that, except for a couple of mines, its isolation meant that it was little known and also the mining period and the period of popular photography did not overlap for long. The mines came and went and the area quickly returned to its agricultural nature. In another way this has been fortunate since much has survived from the old mines; the ruined engine houses, decaying chimneys, adits, shafts and tips remain and over the last twenty years there has been a policy of conservation on the part of the local authorities. Most of this has been successful and has retained for our generation and future generations some of the feel and excitement of the place.

This book provides an opportunity to put back life into the ruins by comparing the present to the old photographs and other images. Fewer than twenty per cent of the photographs were reproduced in the previous work *Mines of Shropshire*, produced a quarter of a century ago (Moorland Publishing, 1976). The older photographs have been joined by images of the 1960s – now 'old' themselves, often showing a very different condition to their present post-conservation state. Buildings have been cleared of their vegetation, consolidated or have even disappeared. Some of the larger mines are beginning a new life as monuments to the past and sites of educational value.

Every care has been taken to cross-check the information given with the images but, as so much has depended on human memory, it is possible that some information is not as accurate as that obtained from documentary sources. If anything is wrong the author must take the responsibility and apologises in advance.

The grid reference number of most of the sites is given in the mines index to enable the location to be identified on the local Ordnance Survey Maps, copies of which would be useful to the reader in locating the former mine sites.

Many structures remain but most of the artefacts have long disappeared and it is hoped that the remainder can be preserved. Some have been removed for safety or conservation but it is anticipated that they will be available for inspection one day.

The Shropshire Mines' Preservation Trust, a charitable trust set up in 1996, is very active in forwarding these aims. It has trustees from both statutory and voluntary bodies and already some sites have been opened up with trails and interpretative plaques, and an Information Centre has been opened at the Snailbeach Mine.

Shropshire Minerals 'Spar Box' *c*.1900, made of crystals, rocks and minerals found in local mines. This is the only known example of a locally made 'box' assembled by a local miner in the late nineteenth century. In some other British orefields, principally the North Pennines, they were quite common. Clearly seen in this box are crystals of calcite, quartz and witherite, other 'sparkles' on dark surfaces are crystals of the ores of lead and zinc. The box itself would have been fashioned in the taste of its period probably for a hall or 'front' room.

One

The South West Shropshire Orefield

In comparison with other British orefields the Shropshire field is small and compact, lying roughly within a radius of six miles around Shelve village, thirteen miles south-west of Shrewsbury. Despite its smallness the area has a long and interesting history; pigs of lead found locally bearing the stamp of Emperor Hadrian attest to Roman working while prospecting for minerals has continued even into recent years. The chief minerals sought, roughly in order of importance, have been the ores of lead, zinc, witherite, barytes and calcite together with a little fluorspar and copper. The mineralisation occurs mainly in Ordivician strata, although some is found in the pre-Cambrian rocks in the eastern part of the orefield. The distribution of the ores suggests some form of arched depth-zoning, the outer zone being barytes and witherite, below which lies lead, then zinc. The crown of the arch has been eroded away so that the deeper zone minerals now outcrop in the centre of the field west of the quartzitic Stiperstones edge, while the barytes outcrops on both the eastern and western flanks of the orefield. The most important mineral producer in the orefield was the Snailbeach Mine, but other highly significant workings were the complexes based on Romans Gravels, Tankerville, The Bog and the Grits. In the present century the mines around Wotherton and Huglith have been foremost in the production of barytes, extraction ceasing as recently as 1948 at the latter.

The mineral ores are found in veins which cut through the native strata of the area; the veins are generally steep, often near vertical, and may vary considerably in thickness being even up to 20ft or more. This is rare, however, for most of their length they are much thinner than this. Around Shelve the mineral veins are found in strata of both pre-Cambrian and Ordovician age. The main products have been the ores of lead and zinc, barytes, witherite, calcite and fluorspar from veins in the Ordovician but barytes has been the chief mineral produced from the pre-Cambrian. Occasionally other minerals have been found and worked in the area including ochre and copper, and silver has also been extracted from the lead ore.

Some of the earliest geological research on the area was carried out by the eminent geologists R.I. Murchison and G.H. Morton who published papers in 1839 and 1869 respectively. Murchison spent considerable time in the area and his work was included in a two volume book on the Silurian System. Morton produced his paper in two parts as lectures to the Liverpool Geological Society. It was later to form a small thirty-nine page book. Other important geologists added to the total knowledge later in the nineteenth and early twentieth centuries. The descriptions given indicate that, at least up to the middle of the nineteenth century, the mining methods were rather primitive, the workings were wet and that the shafts were fairly deep. Aikin (another geologist) said that Snailbeach was 180yds by 1797.

The veins are found in hilly areas, the Stiperstones rise to over 1,700ft AOD, and this is helpful in both finding them and in their working. Water, as rain and in streams, can erode the surface cover to expose the veins while any shallow excavation into them can be drained by tunnels or adits at a lower level. The altitude has also assisted in providing for ventilation and the draught necessary to operate early lead smelting techniques. William Hooson in 1747 described how the draught assisted the melting of the ore in 'boles' 'upon the hills called Stiperstones in Shropshire'. He said 'they are very little places' made by 'laying a round row of stones on the ground, and placing the fire in the middle'. It is mentioned that perhaps only one

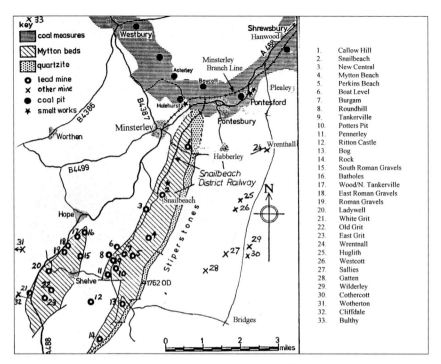

Some of the mines and smeltmills of the Southwest Shropshire orefield and part of the Shrewsbury Coalfield.

key
- coal measures
- Mytton beds
- quartzite
- ○ lead mine
- × other mine
- ● coal pit
- ✱ smelt works

1. Callow Hill
2. Snailbeach
3. New Central
4. Mytton Beach
5. Perkins Beach
6. Boat Level
7. Burgam
8. Roundhill
9. Tankerville
10. Potters Pit
11. Pennerley
12. Ritton Castle
13. Bog
14. Rock
15. South Roman Gravels
16. Batholes
17. Wood/N. Tankerville
18. East Roman Gravels
19. Roman Gravels
20. Ladywell
21. White Grit
22. Old Grit
23. East Grit
24. Wrentnall
25. Huglith
26. Westcott
27. Sallies
28. Gatten
29. Wilderley
30. Cothercott
31. Wotherton
32. Cliffdale
33. Bulthy

hundredweight of ore was smelted in any one place. The early methods of working and smelting mentioned above probably had not changed since the time when the first miners worked the area. The earliest known are the Romans who left evidence in the form of ingots of lead and tools – although as yet there is no conclusive proof of the latter.

Early Workings.

It is believed, and there is some evidence for this, that the Romans had a fort in the area and that they, or those working for them, used water channelled across the hillsides to locate the veins of mineral. This was then quarried down from the surface to remove the ore and the waste using primitive wood shovels, and possibly baskets for carriage. The excavators of a villa site near Linley held the view that it had been used by persons involved in mining and regular channels for water can still be seen on local hillsides. The evidence is much more certain for Roman work than other early working, when it is considered that three pigs of lead with Roman inscriptions have been found in the county. These were found at Aston, north west of Bishops Castle, at Snailbeach and at the Roveries, near Snead. It has been claimed that a pig was found at Roman Gravels Mine but no real evidence for this is available. All the Shropshire pigs have the same inscription which is understood to mean 'the property of the Emperor Hadrian Augustus'. All three pigs are said to carry another mark, that of a palm branch, but the meaning of this is unknown. The historian Thomas Wright also recorded that Roman coins and pottery 'have been found from time to time among the old rubbish' at the Roman Gravels Mine.

A study made in 1943 of the location of the pigs of lead found in Shropshire gave the following information;

1. found at Snailbeach in 1796, deposited in British Museum in 1798.
2. found at Snead in 1851 while draining a field, acquired by a Joseph Mayer and subsequently presented to Liverpool Corporation.
3. found near Aston Farm in 1767, later seen in Linley Hall by Morton and Wright, then lodged in Birmingham Geological Museum.

A Roman 'pig of lead' and early spades from the Shropshire mines. The pig of lead is probably the one from Aston Farm. The pigs had dimensions of about 19in length and 7in width, they weighed 170-200lb. The spades were said to have been found in old workings at Shelve. They are of oak, about 9in wide and 12-16in in length. The centre hole was apparently for the insertion of a handle. (From *Roman Mining in Britain* by G.C.Whittick, Transactions of the Newcomen Society, Vol.XII, 1931-2).

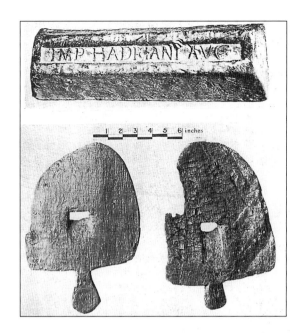

There is some certainty that spade-like tools have been found in the old workings at Roman Gravels Mine, these have been photographed and they are interesting because, unlike modern spades, they are of wood and have a hole for a handle in the middle of the blades. (Another group of wooden mining tools is still on display in the library of Shrewsbury School but this group is believed to have been found in the Llanymynech Copper Mine, in North Shropshire).

From the eighteenth century.

Very little is known of any mining in the area until the mid-eighteenth century although there must have been some small-scale working. A map of the area was produced by Rocque in about 1750 and this shows 'Bog Mines' and 'Penaly Lead Mines' clearly, while Baugh's map of 1808 shows 'Bog Mine', 'Gravels Mine', 'White Grit Mine' and 'Snalbach Mine Works'. 'Penaly' and 'Penally' are early names for Pennerley while 'Snalbach' is obviously an early name for Snailbeach. Despite the lack of information about Snailbeach on Roche's map this mine certainly existed in the 1750s since it is shown on estate maps of the time and an inventory, even showing an 'engine', is available for 1769.

The population in the mid-eighteenth century must have been very small for few houses are shown on the estate maps of the time and, even in 1851, Bagshaw wrote in his Directory that Minsterley, the local market town, 'half a century ago... consisted of only a few farm houses' and that, between 1811 and 1831, the population of Minsterley parish rose from 705 to 809.

The earliest large-scale mining attempts seem to have been made at the south end of the Stiperstones Hill. The workings at the Grit or 'Gritt' came to the fore with the activities of John Lawrence, and later his son of the same name – local landowners between 1780 and 1830. As well as having an interest in most of the mines in the 'Hills' area, except for Snailbeach, they were heavily involved in driving two large drainage levels, the Hope Adit and the Boat Level. They were also interested in steam engines, and during their time several were brought into the area. Snailbeach Mine, although it had operated since at least the early eighteenth century, didn't come to be important until the 1780s.

At their peak the Lawrences controlled seven lead mines, as well as two smelt mills and four collieries in the nearby Shrewsbury Coalfield. The Lawrence empire (and their White Grit Co.) ran into difficulties with other local landlords, on whom it depended, in the 1820s and by the 1830s lawsuits had cost the Lawrences so much they were forced to sell many of their posses-

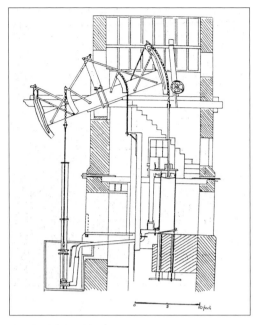

Steam engine for the Grit Mines 1783. A sketch of a drawing in the Boulton & Watt Collection, Birmingham. An inventory of 1767 shows an engine, possibly a steam engine, at Snailbeach Mine, however between 1777 and 1793 at least eight Boulton & Watt engines were brought into the area. A note on a map of about 1860 says 'there are 17 steam engines in the Rea Valley' alone and four others are on order. The remains of the steam engine houses are an important feature of the area today. (SCMC Collection).

sions. This included ground the family had held for over 200 years. The other two major landowners in the area were the Mores of Linley Hall and the Tankervilles.

To the north of the orefield the Snailbeach mine operations were more successful in the late eighteenth and early nineteenth centuries under the Lovett family. The mine was on the land of the Marquis of Bath with an adjoining mineralised area on the Vessons estate.

The other major mining area around The Bog was owned by the Lister family, Lords of Rowton. The Bog was an old mine which came to the fore in the late eighteenth century, first when steam engines were erected in 1777 and 1789, and then, as one of the mines in the Lawrence empire supplying ore to the smelters at Malehurst and Pontesford.

During the 1830s and 1840s the price of lead was low and little investment was made in the mines – indeed, many had closed. By 1850, however, an upsurge had occurred and considerable new interest developed. New companies became involved and the mines attracted miners from a wide area. Some new engines were erected but by the end of the decade the price was falling again and most mines closed.

Although ore prices were still low in the 1860s the mines had been returned to local enterprise and, fortunately, rich deposits were found at Tankerville and Roman Gravels. Snailbeach, too, continued to produce good returns despite low ore prices but this was partly because of good management introduced from elsewhere, particularly by the Eddys.

The 1870s saw the peak years for the area and production of ore rose to nearly 8,000 tons lead ore per year in 1875. This was the period of new and bigger engines, of increasing population and housebuilding and of the strong religious feeling that led to the opening and re-development of many local chapels and churches.

Working the mines

By the 1850s the mining field had been divided up into 'setts'; each sett or area being allocated to a different mine. How and when these divisions were made is not known but they were certainly well established by this time. These names did change sometimes, often to make the association with a neighbouring sucessful mine closer – and thus, to make them more attractive to speculators. They were also sometimes divided or combined. Setts were taken on lease from the mineral owners and companies were promoted to raise the necessary capital for development. Shares were bought and

Seventh Earl of Tankerville and Lady Leonora, c.1907. The Earls of Tankerville were major landowners in the area and keen to take advantage of its mineral wealth. The 7th Earl stayed in the area for a short time about 1907 trying to promote mining and provide work for the local miners. During his time in Shropshire he handed out photographs of himself and his wife, Lady Leonora, in Coronation robes. They were associated with evangelists and held services in chapels and the open air. Lord Tankerville was instrumental in reopening Roundhill Mine at this time.

an agent and secretary appointed. These would then be supported by managers, or 'captains', for surface and underground operations. These captains were brought in from other mining fields; North Wales, Yorkshire or, more often, Cornwall. The mine operators might also employ engineers.

Miners were recruited locally, and others came in from further afield either lodging locally, sleeping in outbuildings, or, if they thought prospects were good, using the traditional local 'squatters' rights'. A good description of this survives where miners at Perkins Beach would 'select a site on the mountain', obtain assistance from his fellows, purchase some larch cuttings from a nearby plantation, then bring them to the site one day and get them erected to form the shape of a hovel. At the same time other friends would dig sods to form walls and a light roof which would be covered with 'straw, generally mixed with heather' to ensure that it was waterproof.

The miner's family would move in before morning, when usually the farmer, 'terribly chagrined at the damage done to his sheep run', would arrive. Following 'hard after him would come the Lord of the Manor or his Agent to demand a nominal rent'. From the contemporary descriptions this appears to be all part of some overnight charade – for the Lord needed miners to make his royalties and would eventually get a house for renting at someone else's expense. The description continues, 'in most cases' the Lord would then 'grant an allotment adjoining' and soon it would become a 'convenient dwelling' with a garden and a cow or two and as time went by the original structure would be replaced by one more permanent.

New 'mines' were sought by searching eroded ground for likely rocks or indication of veins, by making trial adits and shafts or by re-opening old workings. Following the direction of the veins at existing workings was perhaps the most usual method but occasionally luck is also of assistance. The finding of the vein at Perkins Beach is said to have occurred by a miner 'ditching' around his newly acquired allotment.

The driving of adits and sinking of shafts and associated tunnels proves the existence, or otherwise, of the vein, and the adit or shaft can then be used for access, ventilation, transport and drainage. The vein would be opened out by driving parallel tunnels one above the other at set

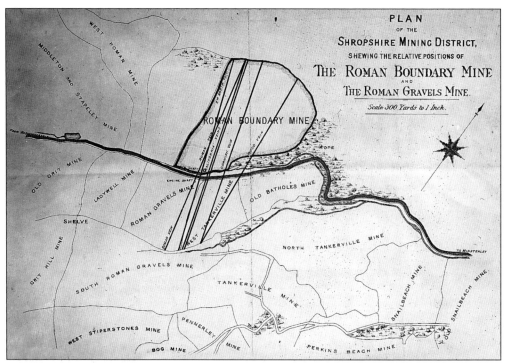

PLAN
OF THE
SHROPSHIRE MINING DISTRICT,
SHEWING THE RELATIVE POSITIONS OF
THE ROMAN BOUNDARY MINE
AND
THE ROMAN GRAVELS MINE.
Scale 300 Yards to 1 Inch.

Mine promotional literature Roman Boundary Mine Sett c.1870. From a handout used to promote the pro-ject, an 'extension' of the rich Roman Gravels Mine. The whole area was divided into mine areas or 'setts' and the plan shows how it was expected that the veins of mineral would continue into the proposed new area.

intervals, connecting them with underground shafts, and then opening these out in the vein. This 'opening out' could be done by removing the mineral from below the worker (underhand stoping) or from above him (overhand stoping). This broken mineral could then be passed down a chute to the haulageway beneath and taken to the nearest shaft or adit mouth. Transport underground was usually by wheelbarrow or small truck on rails. The loads were then raised through shafts in boxes or cages or taken to the adit mouth pushed by hand or drawn by haulage engine. Raising loads through a shaft (called winding) could be done by a horse gin or, in later years, steam or electric power. Drainage from the workings was often done manually in buckets or by simple pump but in many cases at shallow depths it could be done by a purposely constructed adit from a nearby valley. There were many small drainage adits in the area but also four larger ones Hope or Wood Level, Leigh Level, Snailbeach Drainage Level and Boat Level. This latter level also served as a route for the conveyance of ore by boat hence the name. According to local newspapers the boats were sold in 1830. From 1777 Watt's steam pumping engines were being brought into the area for drainage purposes and, from the 1830s, Cornish and other engines were introduced. At Snailbeach a waterwheel was used at the mouth of the drainage level to provide the motion through a rod system to operate pumps in the main shaft nearly a mile away. This was replaced by a Cornish Engine in the main shaft in the mid-nineteenth century.

From the 1870s steam winding engines replaced the manual and horse winding equipment at many of the deeper shafts, a few of the steam engines were dual purpose, used for winding materials and ore as well as pumping. Until the 1870s there were no cages in which men could travel through the shafts although boxes and kibbles (large buckets) could be used. The usual method of travel was to climb timber ladderways built into the shafts. One miner, Samuel Jones of Snailbeach, informed the Kinaird Commission in 1863 that it took him about thirty minutes

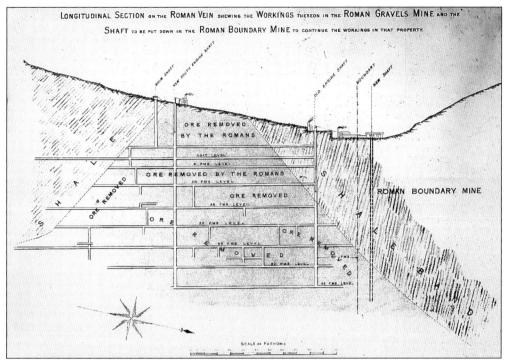

Mine Section as shown in promotional literature c.1870 The layout of the Roman Gravel mine as shown in the literature for the extension into the proposed Boundary Mine (right). The section shows clearly how a vein work-ing was opened by shafts and levels with 'stoping' between. Such prospectuses had always to be believed with caution. The mine sections drawn for Shropshire metal mines frequently showed a 'thumb-nail' sketch of the actual surface plant, horse gins, engines and dressing sheds.

to climb down to his work place at 372 yards each day and one hour to climb out.

The ventilation of the mines was normally done naturally, the air passing down shafts and through adits in the direction depending on the season and wind direction. Tunnels with only one entrance sometimes had to be ventilated by other means, in 1863 the Kinaird Commission was told that manually operated 'fan machines' were used or occasionally 'pipes communicating with a waterfall'.

In the mid-nineteenth century the rock and ore was 'stoped' out using blasting, holes were drilled by hand and then charged with explosive. The firing of these explosives caused much fumes and miners often had to wait over an hour for the fumes to clear. There was some concern about the dangers of these fumes to health.

There were two types of miners underground, 'tutworkers' and 'tributers'. 'Tutworkers' were those who worked on non-ore getting tasks such as shaft sinking, driving tunnels and haulage who were often paid by distance gained – so much per yard. The tributers worked the ore from the stopes and were paid on the amount and sometimes quality of the ore obtained – at so much per ton delivered at surface. The miners worked between six and eight hours per shift, depending on agreement with the company, and of this about one hour might be spent getting to and from the underground workplace with up to half an hour break for food. The men would work in groups or 'tributes' under a leader and he would share the earnings, either fortnightly or monthly, the share-out taking place on a Saturday at the mine or in a nearby inn. The share was either in cash or as a 'note', in either case it would have to be converted into small change. The use of the inns was therefore an obvious but unfortunate location for the share-out and some local mining companies did use them e.g. Roman Gravels. The larger companies also arranged 'stoppages' in the earnings for candles, explosives and a Sick Club.

The managing director and 'Company's Ford Car' 1917, from the prospectus of Wilderley Copper Mine. From the 1860s photographs have occasionally been used in prospectuses for mine development (see photographs of Central Snailbeach Mine). Here, an example from the most recent – a collection used when promoting a mine in 1917. Its original caption read simply 'Managing Director – Company's Ford Car'. Other photographs from this prospectus are shown on pages 102-4. This prospect did not develop into a successful mine but employed eighteen men from about 1917 to 1920. Ore in commercial quantities was not located. (IGMT).

The Mine Surface.

The most obvious features at many of the mines were the steam engine houses for winding and pumping but there were also steam engines for operating dressing plant and latterly air compressors. Other 'engines' included waterwheels and gas engines. Most mines had a cabin, changing house or miners' dry, sometimes small but occasionally large. The changing house at Snailbeach in the 1860s was 'a row of sheds about 70ft long by 10ft wide, divided into 6 houses'. They were very low with no ventilation, very dusty and with earth floors. There was no provision for washing. In the following decade a more substantial building, the miners' dry was provided. Other buildings might include boiler houses, workshops, stores, an office, explosives magazine and an ore house with or without a weighbridge. Snailbeach was an unusually complete mine as it had all these components plus a candle house or factory and its own railway system with locomotive shed and a smelt house. Snailbeach also had a count-house (a more impressive office), its own sawmill and several dressing floors with associated buildings. Larger mines usually had a reservoir or two and a narrow-gauge mine-railway system.

The dressing plant at a mine often consisted of screens, crusher, washing equipment, jigs and buddles and many of these processes required the use of water either for separating materials or for driving water wheels.

The orefield is estimated to have produced about 237,000 tons of lead concentrate and 21,000 tons of zinc concentrate, and has a recorded production of 565,000 tons of barytes and 950 tons of witherite.

The Shropshire lead mines did not employ young boys (under ten years of age) and most, particularly Snailbeach, did not employ women. No record has yet been found of any mine employing more than six women.

William Oldfield (1851-1928) the last 'mine captain' at Snailbeach Mine. Captain Oldfield, unlike many of the other Shropshire mine captains who came from Cornwall, was born at Wrexham in 1851. He came from a mining family and, from the age of fifteen to thirty-nine, worked at Minera Lead Mine. He was transferred by, Henry Dennis, who had interest in both mines, to Snailbeach in 1880 and remained a manager for thirty-eight years. He was given the traditional title of captain out of respect. He remained in charge of operations at the mine even after the main closures in 1912. Captain Oldfield also invested in other mines and worked as a consultant. His collection of minerals was most impressive and was given to Birmingham University and the British Museum. On his memorial in Pontesbury Congregational Churchyard his wife included the words 'Treasured Memories, my husband Capt. W. Oldfield, West View, Minsterley, who entered his heavenly rest Feb. 7th 1928'.

In 1895 Captain Oldfield played an important part in the recovery operations after the Snailbeach Mine Disaster and after the mine was sold in 1912 he became manager of the Halvans Lime Spar & Concentrating Works which took over the Snailbeach site. He remained in this position until his death aged seventy-seven. He was at one time also the manager of Snailbeach District Railway and for, a short period, Bog Barytes Mine. (Mrs D. Lewis, Condover).

M.. *190*

Dr. to WILLIAM OLDFIELD.

Terms : *Monthly Payments.*   £   s.   d.

Captain Oldfield's invoice heading.

A Snailbeach Band, early 1900s. Most events in Snailbeach used a 'band', for example, at the opening of the smelter in 1862 a procession of 400 men, preceded by a band, marched from Minsterley to the new smelt house (about four miles). In 1892, and in other years, the Oddfellows (Earl of Tankerville's Lodge) assembled at Tankerville Mine and walked to Snailbeach Chapel preceded by a band and after the service walked back again for lunch (about two miles each way). In 1892 the parade was headed by Captains' Waters, (Roman Gravels), Job and Oldfield (Snailbeach Mine), Smytham (Tankerville and the Bog Mines) and Mr Toye (Perkins Beach Mine). The Captain Waters mentioned was Arthur Waters, son of the late Captain Arthur Waters of Tankerville Mine. (Mrs S. Hartshorne, 'Never on Sunday' project).

The headmaster and his wife, Stiperstones School 1875-1900. There was a school in each of the mining centres, for example the Stiperstones, The Bog and Shelve. The school at nearby Minsterley was built at 'the joint expense of the Marquis of Bath and Other Gentlemen of the Snailbeach Company' about 1840. The company endowed the school with £40 yearly and ordered that every miner should pay the schoolmaster sixpence per quarter. John William Moore 1843-1916 was headmaster of the Stiperstones School (built 1872) for twenty-five years – he is also believed to be the J.W. Moore who was secretary of the Oddfellows Friendly Society to which many of the miners belonged. The school, too, was more than a school as it served also as a community centre. (Madeley Rest Room Review).

A sketch of the winding arrangements at Snailbeach Mine 1895. This sketch, from the memory of the old miners, was drawn in 1963. It shows the Snailbeach drainage adit, the main pumping and winding shaft from the 342-yard Level and the man-winding arrangements from 252-yard Level at Old Shaft, complete with headframe and enginehouse. From the foot of Old Shaft, the miners climbed down three ladderways, each of 90ft, and then along the 342-yard Level, then travelled down an inclined shaft (with a compressed air winding engine) to the bottom of the mine at 552 yards depth.

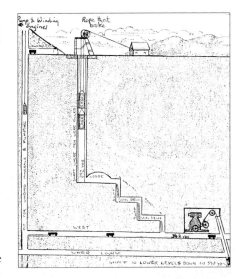

On the fateful morning of Wednesday 6 March 1895 seven men were being lowered down Old Shaft by engineman George Williams (twenty-one years at the mine to this date and hence the common name George's Shaft for Old Shaft) when halfway down the rope broke and their cage crashed to the bottom. The scene was horrendous; all seven died, three of the victims were well-loved local preachers, all except one of the victims left children (John Purslow, aged fifty-two, left ten children). There was great distress in the small community. There was a further death – Anne Blower, another miner's wife, was informed incorrectly that her husband, John, was on the cage which crashed – she collapsed and died the following day. Another wife miscarried her only son, she already had three daughters.

The verdict of the Coroner's Jury was accidental death; it was agreed that the company and chief agents should be censured for allowing a rope to be used for eight and a half years without taking special measures to ascertain that it was safe. But no law had been broken and it took another forty years before legislation was approved which limited winding rope life to three and a half years with regular re-capping.

Two old mineworkers who remembered the 1895 disaster, 1963. Left: Alfred Hewitt, born 1878, was on the way to work when he saw men stretching out the broken rope. His father had been blinded in a mine accident in 1883 (see page 36). Alfred went on to operate the Snailbeach Cornish Engine for £1 per week and after the mine closed he worked at the nearby barytes mine. When he died in 1974 he was said to be 'the last true Shropshire lead mine worker'. Right: Samuel Morris, born 1883, remembered the accident and that his school was used as a mortuary. He started work at Snailbeach as a surface worker but when the mine closed he worked underground at a local barytes mine – and, like others, fell victim to silicosis, a dust disease. (*Shropshire Magazine*).

Above: Mourning card for victims of Snailbeach
Disaster 1895. *Below:* Some notes 'In Memoriam'
from the Primitive Methodist Circuit 'Plan'
showing the appointment of preachers in the
Minsterley Circuit prepared by Revd J. Cope for
April-June 1895. The Plan shows how active the
circuit was with sixteen chapels and meeting
places served by one minister (Revd Cope), sixty
'local preachers', four 'prayer leaders' and
seventeen 'helpers'. It also advertised midweek
services, tea meetings and camp (open-air)
meetings. For comparison, a Minsterley Circuit
Plan of 1993 listed nine chapels served by one
minister and seven 'local preachers'. In 1993 far
fewer services were being held at each chapel.

Right: Front of Memorial Booklet produced by Revd
J. Cope. A twenty-eight page Memorial Booklet
was produced by the Primitive Methodist Minister,
Revd J. Cope, giving brief details of the incident
and of the victims, most of whom were closely
involved in church and village life. It also contains
many poems of varying length and nine hymns.

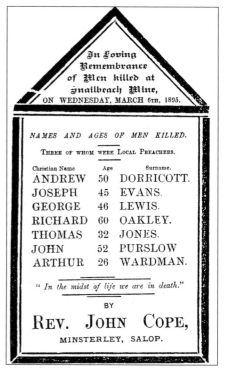

In Loving
Remembrance
of Men killed at
Snailbeach Mine,
ON WEDNESDAY, MARCH 6TH, 1895,

NAMES AND AGES OF MEN KILLED.

THREE OF WHOM WERE LOCAL PREACHERS.

Christian Name	Age	Surname.
ANDREW	50	DORRICOTT.
JOSEPH	45	EVANS.
GEORGE	46	LEWIS.
RICHARD	60	OAKLEY.
THOMAS	32	JONES.
JOHN	52	PURSLOW
ARTHUR	26	WARDMAN.

" *In the midst of life we are in death.*"

BY

REV. JOHN COPE,
MINSTERLEY, SALOP.

Left: The Edward Medal as presented to two Bulthy Miners 1921. The medal shows the head of the monarch on the front and on the reverse (shown) the rescue of a miner and the legend 'For Courage'. *Right:* A 'bogey' in a mine tunnel 1989. This bogey was found in a tunnel at Snailbeach Mine as it had been left probably over seventy years ago. The difficulties that the miners had in rescuing their colleague can easily be appreciated. The modern explorer shown is Peter Ward. (Kelvin Lake/IA Recordings).

Very few non-coal miners in Britain have been awarded the Edward Medal (the Miners' VC) for gallant attempts at saving lives in mines. For their efforts on 29 April 1921 two miners at the English Barytes & Mining Co. Ltd's Bulthy Mine near Middleton, Shropshire, were each awarded the Edward Medal in Bronze for 'together saving a foreman overcome by petrol fumes'. The miners were Robert Edward Paddock and Joseph Thomas Chidley and the foreman was Joseph Henry Bennett. There were at the time of the incident two other miners underground, William Grant and William Whittall. Grant died but Whittall eventually got out alive.

The mine consisted of a shaft and some tunnels leading off, air was being supplied by pipes from a compressor on the surface. A petrol engine was installed underground for pumping. Paddock when walking in met Bennett coming out, Bennett collapsed and Paddock took him through fumes and smoke to the shaft bottom. Chidley arrived and took Bennett to the surface then Chidley and Paddock returned into the mine to look for two other men, Grant and Whittall. They found Grant collapsed, 170 yards in, put him on a 'bogey' and started pushing him out. Paddock himself fell ill and had to struggle out, Chidley found he could no longer push the bogey and tried to follow Paddock.

Paddock had got to surface, reported to Bennett, who then went down and fetched out Chidley but then collapsed. While on the surface Bennett had sent for a youth named Pryce to help. Pryce arrived, went down for Grant but could not get the bogey back on the rails – he collapsed. Two farmers arrived, went down and brought out the collapsed Pryce. A doctor arrived and went with Bennett into the mine to try to revive Grant but they failed. By now, they all thought that Whittall, the missing man, must be dead.

Soon after, the shaft signalling bell was heard and the kibble (large bucket) on the winding rope was raised, Whittall was found alive in it. Whittall explained that he had left his place of work, gone to stop the petrol engine and on the way out collapsed, fortunately, by the end of the pipe discharging the air from the compressor. This saved his life.

The owner and agent for the mine were later prosecuted for operating a petrol engine underground, Chidley and Paddock were awarded the Edward Medal. Bulthy Mine was right on the border between England and Wales, it had been opened in the 1870s and employed up to fifteen workers during the First World War. It closed around 1922.

Cover of the Memoir of Samuel Hughes, *A Shropshire Miner*, 1878. This memoir was edited by W. Benson and published in the year of Hughes's death. It contains twelve pages of memoirs, fifty-one pages of hymns and two long letters. Samuel Hughes was born in 1809 at Habberley and at the age of twelve started work at Snailbeach Mine. He said that during his youth he 'gave loose to the reins of folly and wickedness; the wake, the fair, the races and sometimes even fights'. At the age of twenty-two he married and started a family, then one day he saw the minister and other persons building Lords Hill Baptist Chapel (in 1833 or 1834). This lead to a religious experience which changed his life and he became a local preacher and hymnwriter. He continued at the mine and built a cottage for his family in Crows Nest Dingle. As he got older, he left Snailbeach for the North Wales mines where he got lighter work for ten years. He then returned to his Crows Nest cottage and in 1878 died there. Many of his hymns are sad, particularly one written in 1848 when two of his children died on the same day. The biblical inscription is from Psalms 119 verse 130.

Lords Hill Baptist Chapel 1963. The church and chapel played a major part in the life of many miners in this area and Lords Hill was built in a remote setting high on the hillside above Snailbeach Mine. Mary Webb in her book *Gone to Earth* calls the area 'God's Little Mountain' and the chapel and its minister play a part in it. A film was made of the book in 1952. The first chapel on the site was opened in 1835 but this was rebuilt in 1873 to appear as it does above. For many years its members have worshipped at the mine in the winter months where they have a Sunday school in the old mine ore-house. The grave of Arthur Wardman who was killed in the mine disaster lies just inside the churchyard here. Snailbeach Mine's Chapel Shaft is just off the photograph to the left and Yew Tree Level is hidden by the vegetation.

Minsterley Fair about 1900. The road junction in the centre of Minsterley, in front of the Miners' Arms Inn (now the Bridge Inn) was for many years the site of markets and fairs. This postcard shows one of the fairs with locals and traders milling around. Just off the photograph were three other inns, a chapel and the Anglican Parish Church with several memorials to agents and mining families. Over a couple of fields to the right of this photograph is the Asterley Windmill now restored and with sails again.

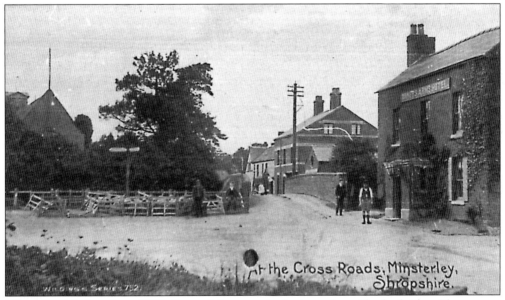

Minsterley Market about 1900. Pens of sheep are to be seen alongside the bridge. The present bridge was opened in 1910, and with the closure of the local mines, the adjoining inn has been renamed 'The Bridge Inn'. Close to where this photograph was taken there is still a hole in the wall leading to a useable cast-iron gents urinal, a remnant of days when the miners and others congregated here.

Postcard of Snailbeach c.1910. Several postcards exist showing the attractions of the area; there have been many visitors over the years. These early tourists often came to visit relatives who had moved here with the various tides of success at the mines, as well as geological students. Of particular interest in this card are the views of the mine workings on Lords Hill Bank (top left), the long closed Cross Guns Pub at Crows Nest, the nearest inn for the Snailbeach mines (centre), and an inside view of the now closed 'Cosy Cafe' (bottom right).

Party of visiting mining history enthusiasts 1965. One of many hundred such parties coming to study the mining remains. Passing in front of the old compressor house this one contains three people, Ivor Brown, Kenneth Lock and Richard Haszard, who have each been leading mine study groups around for forty years or more. Interest in these remains is as great, if not greater, today than ever before, assisted by such bodies as the Shropshire Mines' Preservation Trust.

Snailbeach Mine Cornish Pumping Engine House, 1963. This large engine house has been a prominent feature on the hillside overlooking the village for about 150 years – it can be seen for miles around. Built in the 1850s it housed a 60in Cornish engine to power the pumps for draining the mine. The deeper workings were abandoned in 1911 when pumping ceased.

Snailbeach Mine Cornish Pumping Engine House, 1994. During the thirty years after 1963 vegetation took a stranglehold on the building and the roof eventually collapsed but in 1993/4 conservation and consolidation work was undertaken. The large opening for the beam, the holes for the floor supports and the front window became clear again, as did the shaft. Unfortunately the highly visible roof outline was not maintained and the straight roof with twin points on the rear wall now gives it the appearance, from a distance, more of a factory roof that what it really is.

Badger exiting from mine working 1960. The area has long been known as a haven for wildlife – part of it is now a National Nature Reserve and a larger area is designated an 'Area of Outstanding Natural Beauty'. Badgers have been common here, using the old mine workings – as shown when the author met this one – when he was entering the mine just as the badger was coming out.

Zinc mine plan angers residents

The country's only zinc mine could be opened in the Hope Valley if planned test boring proves successful.

But angry residents, who claim they have been kept in the dark about the proposed ten year mining project, are now calling for a public meeting with the company behind the plan.

The company, TE Minerals, is part of the Tankerville

Shrewsbury & Atcham District
Application No S92/1141

Planning Commi
10 December 19

PROPOSAL IN RELATION TO MINERALS EXPLORATION –
COTHERCOTT BARYTES PROJECT. TRENCHING TO ESTABLISH
THE CONTINUITY AND WIDTH OF BARYTES MINERALISATION TO THE
WEST OF THE OLD COTHERCOTT BARYTES MINE, PULVERBATCH –
BAKER SILLAVAN LTD.

Newspaper and other cuttings 1980s – 1990s The hope for mining again is not yet dead: these cuttings show that the search for minerals continues. Tankerville Estates have been active in recent years in assessing their reserves of zinc and lead, and trying to sell the mineral rights. Other companies have searched for barytes and other minerals. One exploration company still has its base locally but most of its work is done further afield.

Two

The Snailbeach Mines

Snailbeach Mine was certainly the richest lead mine in Shropshire, producing 131,900 tons of lead ore between 1845 and 1913, over half the lead ore produced in the area during that period, as well as other minerals. It is considered by many to have been, acre for acre, one of the richest mines in Europe. A Roman pig of lead was found here in 1796 suggesting that mining was taking place in those times but documentary evidence exists from the sixteenth and seventeenth centuries.

Extensive working is known to have occurred during the eighteenth century and in 1783 the Snailbeach Partnership was formed. By 1797 mining had reached a depth of 180 yards at Old Shaft and a drainage adit had been driven 1,200 yards from Wagbeach in the Hope Valley in order to assist in draining the mine. A beam engine had been used for pumping to surface but

This is an excellent overall view of Snailbeach Mine, c.1900, one of the richest lead mines of its time. From left to right can be seen, on the hill, the boiler, winding, pumping and winch houses, with the Count House below and beneath that the coal hopper, timber yard and two-storied office. Behind the 'steam' is the compressor house with chimney (1881) alongside, just visible. The crusher house and chimney are to the right of the steam with the dressing floor and sheds in front. Behind and above the sheds can be seen the Old Shaft headframe with engine house (1872) and smoking chimney behind it. To the left is the miners' dry (with roof vents), low workshop and alongside this the day level from which a tramway exits and sweeps round into the crusher house. To the right of the headframe are the blacksmiths' shops, in front of which is the small two storey office overlooking the dressing floor. In front and below the blacksmiths are the double-loco sheds, a train of wagons on the Snailbeach District Railway and below all this the ore-house.

the drainage level allowed a waterwheel to be positioned at the entrance to the level and from there rods powered a pump in the shaft itself.

From 1782 Thomas Lovett was the leading partner and he also controlled a coalmine at Pontesford and a newly erected Smelt House, also at Pontesford. Snailbeach Mine proved to be very rich and by the 1850s was producing up to 3,500 tons of lead ore annually.

In 1857 the company appointed Stephen Eddy from Yorkshire as agent and he carried out a major overhaul of the mine with the aid of his son. One result of this was a substantial reduction in the number of employees and a considerable increase in profits. The pump rod system was replaced by the Cornish Pumping Engine and by 1863 a new smelter was built near the mine to replace that at Pontesford. The company also commenced working zinc ore and barytes.

In 1867 the Snailbeach Mine Co. was formed and by 1870 James Ray Eddy, the son of Stephen Eddy, had resigned as agent and a Cornishman, Henry Dennis, was appointed. A major refurbishment of the mine again took place including a new dressing plant and the large reservoir with a leat system to supply it with water. Old Shaft was deepened and a winding engine erected (datestone 1872), a steam crushing engine was installed and the Snailbeach District Railway narrow-gauge system built. Working was very successful during the 1870s.

In 1881 a compressor was installed and working methods improved but then the market for the products began to fail. By 1896 the Smelt House had been closed but deepening the workings found more ore. The mine reached a depth of 552 yards, the greatest depth in the orefield. In 1895 there was a serious shaft accident in which seven miners were killed. The mine had now become more dependent on zinc ore and barytes production, lead ore production continued to fall. In 1911 the Snailbeach Mine Co. ceased operations but other smaller operators took over various parts of the site.

The mine is said to have employed 500 in 1851 but in 1859 the Mining Journal reported that the Eddy's first act had been to lay off 170 men and boys. In 1877 356 were employed, this fell to 150 in 1900 and below 100 by 1910.

Snailbeach Mine 1900, before pumping ceased in 1911. The 'Snailbeach District Railway' narrow gauge locomotive on the left of the picture is near the foot of the incline which took wagons of coal up to the pumping engine. This railway, which was opened in 1877, can also be seen between the Black Tom shaft headframe and the white cottage to the right of the picture. The side-by-side double building in the centre of the picture is the loco shed. Left of this steam can be seen rising from the complex of buildings which included a winding engine of 1872 and an air compressor of 1881. In front of these buildings is the headframe of Old or George's Shaft. (K.C. Lock).

Snailbeach Mine Yard c.1910.The photograph shows the curved raised tramway by which loaded ore wagons were taken into the crusher house, after being drawn up the shaft. This view was taken looking from a position below the pumping engine on the hill. The compressor house roof can be seen on the right. (IGMT).

Snailbeach Mine Old Shaft Headframe and Yard c.1910. A similar view to the above but showing more of the minebuildings and the tip. The miners' dry is in front of the headframe and the winding engine to the left. The two chimneys to the right are from the compressor house and the crusher house. (K.C. Lock).

Snailbeach Mine Reservoir and Pumping Engine *c.*1910. This photograph was taken looking down the valley from Lords Hill; the Lords Hill adit workings are shown in the foreground. The reservoir was constructed in 1872 to overcome shortages caused by periods of drought. The mine pump on the hill to the left normally pumped only up to the drainage level and this discharged into the Hope Brook, nearly 1,400 yards away. In really dry weather the pump could, however, be used to raise water to the surface. In the foreground are the rail tracks which brought mineral from the Lords Hill Mine workings. (K.C. Lock).

Snailbeach Mine – A group of miners *c.*1900. The photograph shows particularly well the clothing of the time. The usual miner's hat was an old 'bowler' rubbed with resin to toughen and make waterproof. Several such hats appear to have a candle stuck on them with a 'gob of clay'. The candlefat would also assist in hardening the hat. The mine was said to be 'comfortable' and dry. (Emily Griffiths Collection).

Snailbeach Mine Day Level 1993. Ore was raised from the mine up New Engine Shaft (in front of the Cornish Pumping Engine) to a level connecting with this adit at the mine dressing floor. The wagons then came out of this adit traversed a curved tramway and entered the crushing house. This adit or day level has recently been repaired and is accessible to visitors on Open Days and by appointment. Other buildings in this photograph include the Count House (or Agent's Office) above the Day Level and to the left, the 1881 compressor house. *Inset:* Iron Plate formerly on the Day Level. Until about 1965 this iron plate, bearing the inscription 'Snailbeach Mine Co. 1848', was above the entrance to the Day Level, its significance is not known but it could be the date of opening or reconstruction.

Snailbeach Co. Workers *c.*1900. At least two of these men appear to be holding masons' hammers and there is a wheelbarrow of stone to the right. It seems that they might be quarrying stone or constructing stone walls. It is possible that they were working at the company's quarry near the tramway between the mine and the smelthouse, behind Prospect House. The plate '1832' may have come from the old crusher house at the mine which was being reconstructed about 1900. (Snailbeach Methodist Church).

Old Shaft Headframe about 1900. The headframe had pulleys for the two ropes in the narrow 252 yards deep shaft which were used to raise both the miners and some of the barytes during period 1872-1912. At the bottom of this shaft the miners still had a ladder descent of 90 yards followed by a walk along the main tramming level to a further cage in an inclined shaft which took them to the bottom level 552 yards deep. Old Shaft is alternatively called George's Shaft after George Williams, the winding engine-man at the time of the accident on 6 March 1895 in which seven miners were killed when the winding rope snapped. (IGMT).

The Snailbeach Mine.

Old Shaft Winding House c.1965. The winding house stands roofless although its datestone, 1872, can be clearly seen. The engine driver's steps and window for the rope are clear. To the right is the timber headframe of Old Shaft with a single pulley (constructed c.1913) and behind the bushes are the early winding/pumping house and blacksmith's shop. The building on the left of the drawing is the 'miners' dry' or changing room. The boiler house, chimney and reservoir are behind the winding engine house. (M.Newton).

Left: Old Shaft Headframe in 1967. This wooden headframe, with single pulley, was used for winding barytes from above the 112-yard level after the pumps had stopped in January 1911. The iron cage capable of carrying seven men is visible. The headframe continued to deteriorate and collapsed over the shaft in 1974. *Right:* Snailbeach Mine flat-wire winding rope 1994. This flat-wire rope was found during the clearing out of the 'Miners' dry' in 1994. Each strip had been cut to a regular length of about one yard and from their location and marks on the adjoining wall they appear to have been used as treads on the wooden stairway to an upper room. The rope probably dates to pre-1880, it is known that round rope for winding was in use from that time. Flat rope usually consists of up to seven round ropes hand stitched together, a rope installed in 1859 was 3.5in wide by 0.625in thick, 450 yards long and weighed '2 ton 2 cwt 1 qtr 4 lb'.

Snailbeach Mine round wire winding rope 1994. A roll of round wire rope lies in the miners' dry which appears to be similar to that which broke in the mine disaster of 1895. At the time of the disaster the rope was described as 'best steel wire rope, 320yd by 3.5in circumference (1.125in diameter) weighing 16cwt 1qtr 10lb'. Old residents of the village have said that it is the rope which broke and was taken off for complete examination!

*Snailbeach Mine
Minsterley
Salop*

Snailbeach Mine Office June 1967. Prior to the roof collapsing, the office had everything that would be expected of a mine manager's office; a desk with glass ink pots, a flat table for preparing plans, a mine plan fixed to the wall, a corner cupboard, a window ledge with mineral specimens and a fireplace with spare ink cans on the mantelpiece. After the front wall fell the roof went and precious documents were scattered over the site. *Inset*. Mine notepaper heading with simple address, the type on which does not look very old. Other companies on the site from 1900 have also used the shortened form of 'Shropshire' i.e. Salop.

Horizontal Steam Engine near Black Tom Mine 1930s. This engine, which seems to be derelict, is said to have been used for driving a sawmill. All mines had to have a large timber yard and Snailbeach was no exception. (H.F. Wheeler Collection, R.S. Carpenter).

Snailbeach Mine underground, ore wagons 1990. These wagons remain on the 40-yard Level just as they were left over seventy years ago. The timberwork is rotting and they cannot be moved as they are rusted to the rails. (Kelvin Lake/IARecordings).

Snailbeach Mine underground, forty-yard Level 1990. A group of tools; a fork or 'frank', a shovel, shothole stemmers and possibly an oil or powder can along with other implements remain in the mine as they were left when the miners finished work over seventy years ago. (Kelvin Lake/IARecordings).

The Blacksmith's Shop, Snailbeach Mine, about 1900. The long drill steels shown on the truck have been brought from the mine for sharpening. These long steels were used with the compressed air drilling machines which had been introduced to the mine in 1881. By 1882 Snailbeach had four drilling machines in use. Such innovations were not without dangers, and in 1883, John Odgers, 'the foreman of boring machinery' at Snailbeach was killed and several miners seriously injured when the blows of the drill initiated a charge of dynamite. The workmen in the photograph were (left to right) W. Griffiths, T. Jones G. Price and L. Roberts. (K.C. Lock) *Inset:* Mourning Card for John Odgers. (Emily Griffiths Collection).

Members of the Hewitt Family 1890s. The Hewitt family have lived in a cottage on the mine site since at least the mid nineteenth century and their menfolk worked at the mine. Alfred, (1878-1974) rear row, right hand end, was the last surviving 'lead mine worker' when he died. His father, John, born 1835, is the one with eyes covered by a bandage. In 1883 remnants of dynamite in a previously drilled shothole exploded when a new hole was being drilled alongside. The explosion killed one, injured others and damaged Mr Hewitt's eyes to such an extent that forever after he wore the bandage. Young Alfred Hewitt later worked in the pumping engine house and then at other mines in the area. (The Hewitt family).

Snailbeach Mine blacksmith's shop 1967. The photograph shows the forge, anvil and tank of cooling water; at this time the 'shop' was complete with tools but was in a very derelict condition. Nothing much had been done in the 'shop' for about thirty years.

Snailbeach Mine blacksmith's shop 1997. Thirty years later the forge and its ancillary equipment including bellows have been conserved. There is also a single post wooden crane and a full range of tools.

Black Tom Mine, Snailbeach *c*.1912. Part of the Snailbeach Mine, this was an old shaft reopened by the Halvans Co. in 1912 to work barytes. It was sunk deeper and the forty-yard level was further developed. There was a small horizontal engine in a wooden 'house' and work continued until the 1920s. The photograph shows the wooden winding house, and, it seems, the double pulleyed headframe is on its side and about to be raised. The road up to Lords Hill Chapel can be seen above the tips on the right. The Halvans Co. took a sub-lease to dress the tips of the mine in 1900, they erected a steam engine house and plant on the tips themselves. When the Snailbeach Mine closed in 1911 they took over the Mine to work Black Tom shaft and the upper levels at Old Shaft. In 1913 they were producing 3,734 tons barytes using thirty-eight men (twenty-four underground) and by 1918 increased production to 4,634 tons using forty-six men. Thereafter production fell steadily to 745 tons in 1930 when the operations closed. At this time less than ten persons were employed. (M.Davies).

Black Tom Shaft *c*.1912. A photograph taken at the time, 'the last new pithead frame was raised' or 'reared'. The headframe seems to have been constructed on its side and was about to be lifted using ropes. The 'rearing' of a headframe was often a good excuse for celebration. The workers present are, left to right: Fred Blakemore, Jack Corfield, Sam Williams, Alfred Hewitt and Joe Rowson. (Emily Griffiths Collection).

Black Tom Shaft *c.1920*. The group of miners standing in front contains some of those shown at the rearing of the headframe (previous photograph). The small cage and coupling box for the rope can be seen behind the seated miner. (J. Butler).

Left: Black Tom engine house 1967. The engine house was of slatted timber construction with window for the engine driver in the centre. To its left can be seen an opening for the winding rope. *Right:* Black Tom headframe 1967. This has twin pulleys and of very light timber construction because of its shallow depth. The headframe collapsed shortly after this photograph was taken.

Barytes Plant *c*.1930. A group of workers at a barytes separation plant on the Snailbeach site. They are; back row, left to right: Will Hughes (foreman); Billy Pinches; George Adams; Peter Pugh; Joe Rowson; Bob Downes; front row, left to right: Tom Lewis; Tom Jones and Eric Downes. (Brenda Jones).

Working on the Buddles at Snailbeach *c*.1915. This is believed to be Mr Randles controlling the device that separated ore from waste using water. A wheelbarrow is in the foreground. (Emily Griffiths Collection).

Barytes Dressing Plant, Snailbeach 1960. Until recently a simple jig and wooden spiral drum washer stood close to Black Tom Shaft. This was the means by which the Halvans Co. dressed their barytes ore. The run-of-the-mine material would be put over a 'grizzly' (screen), the larger material being handsorted. Small material would fall into a bin and then pass through the rotating drum washer. The spiral would carry the heavier lead and barytes up into the jig where flotation would separate the heavier lead and barytes from the other minerals. The equipment was driven by shafting from an engine.

Snailbeach Dressing Plant c.1940. An earlier view of the plant shown above. It can be seen that it was driven by shafting from an engine to the right, off the photograph. As there is no belt it was probably disused at this time. (Beamish Museum). Inset: Barytes Dressing Plant designed for a Shropshire Mine c.1920. Although this was probably designed for a much bigger plant at The Bog Mine, the same principles apply as above.

Halvans Steam Engine House 1961. Situated near the junction of the mine access road and the road to the village, this engine house erected about 1900, provided power for separating barytes and other minerals on the adjoining tip. The building was demolished about 1985 and this act gave great impetus to the conservation proposals. Action was taken by the County Council the day after the demolition was reported to them, and the proposed demolition of the next structure, believed to have been the compressor house chimney, was thwarted.

Barytes Plant c.1910. Situated opposite to the mine access road where the village hall now stands was this corrugated sheet and timber construction. This site had been a brickworks previous to tipping and by 1896 a barytes plant had been erected. The business does not seem to have been a great success and was closed by 1910. This larger more successful plant was further developed by C.C. Gray in the 1920s. (K.C. Lock).

Small plant on Halvans Yard (site of present Village Hall) c.1900. This may have been the earlier plant referred to in the previous caption. The tips can be seen to be as placed by wagons in 'finger' formation with a small track cut through them. ('Never on Sunday Project').

Spar Separating Plant, Snailbeach 1960. The dressing of spar (quartz and calcite chips) and the production of 'hardcore' had been carried on here alongside the village hall for many years. This trommel and the associated bins were still in place in 1960. The materials were used for pebble dashing and in construction and operations continued on various parts of the tip until the 1980s.

THE GRAVEL TRADING Co. LTD.

WORKS:
SNAILBEACH.
MINSTERLEY.

5 HIGH STREET,
RUABON.

··193·

We have to-day despatched Order No.   *as undernoted :—*

Name and Destination.	Particulars of Sales.	Weight.				Wagon
		T.	C.	Q.	lbs.	Nos.

The Gravel Trading Co. Ltd. The company's despatch note heading. There was a Gravel Trading Co. operating on site from the early 1930s. A local directory of 1941 said that the company's local manager was Mr J. Roberts and that the tip contained 2 million tons of material. A number of men and boys were employed.

Above: J. Roberts. Heading from notepaper.

Left: Perkins or Roberts Level, Lords Hill Mine 1960s. There are at least five levels entering into Lords Hill from the valley near the reservoir, the best known being Perkins or Roberts Level. Perkins Level is recorded in the 1850s, 'Roberts' is a much later description. The name 'J. Roberts' has been found on documents since the late 1920s when the Snailbeach Barytes Co. was supplying Halvan's customers with barytes. In 1944 official records describe Mr Roberts as manager for the Barytes Co., the mine being called officially Lords Hill. It employed four underground, three on the surface and continued to operate until 1950. Work seems to have continued on surface from this date, with a short period of underground activity prior to 1955. It continued to be described as a 'small lead mine' in offical mines inspector's reports until 1961. The Level has recently been made safe internally and is accessible, on advertised days, to the public. Access is controlled by the Shropshire Mines Preservation Trust. This mine shows clearly how a Shropshire mine was worked, the vein is wide and the stopes extensive. (M. Newton).

Ore dressing plant, Snailbeach, 1960. This plant stands outside Perkins Level and was used for the preparation of barytes up to the mid-1950s. A large stockpile of barytes and several kibbles of lead ore can be seen in the foreground. These had been separated from the run of mine material by a simple jigging and screening operation, one of the kibbles of ore is shown on page 4.

Surface roadway collapse, Lords Hill 1963. The mine workings broke through to the surface alongside the roadway on Lords Hill. The photograph shows a tripod erected alongside exposed stope workings. The roadway was diverted around the collapse but, as the affected area extended, a temporary road bridge was put in place. Many other relics of mining could be found along this vein in 1963 including pieces of wire rope, both round and flat, a 'blow george' for ventilating and remains of a hand winch.

The Barytes Vein in Perkins Level 1970. The workings under Lords Hill are quite extensive, and as can be seen the vein is about fifty-five degrees and up to 7ft wide in places. Pillars of white barytes left in to support the hanging wall can also be clearly seen. The vein was worked in this part of the mine as recently as the 1950s. (Shrewsbury Chronicle).

Snailbeach Mine Tips 1967. These operations consisted simply of hand sorting and screening of mineral from the body of the tip, the only machinery normally on site was a tractor with bucket, a static screen and a caravan in use as an office. In 1965 the Snailbeach Barytes Co. notified the Minister of Power that they had produced 8 tons of barytes and 1,020 tons of 'gravel' using 292 man-shifts. The company was still listed as 'hardcore' producers in Quarry Directories in the 1970s.

Snailbeach Tip Land Reclamation 1995. During this year the tip was regraded to make it stable and safe. As the photograph shows, rolls of geotextile material were used to cover the area and soil was spread over this. During excavation work two buddles were exposed. They lie alongside the Snailbeach Railway and were probably late nineteenth century. In the buddles mineral fines in water were put in at the top of a slight slope and, with the assistance of some gentle brushing, the fines would be sorted by weight. After a quantity had built up the separate bands would be dug out and disposed of according to the type of mineral.

Snailbeach/Lords Hill Mine Stabilisation Works 1994. After about fifty years of rest, some of the old workings were reopened and modern machinery brought in to enable stabilisation work to be carried out. This was part of the reclamation and conservation work recommended by consultants. The photograph shows some of the old workings being filled with fine material conveyed down bore holes from the surface. The area treated in this manner was limited for various reasons, including expense, access difficulty and presence of bats, which are a protected species. (Wardell Armstrong).

Snailbeach/Lords Hill Mine Clearance Work 1994. During the reclamation works a 'modern' mini-digger was employed to clear and open up some of the old access tunnels. Part of this area is now accessible to the public. The old miners would have been amazed to see such intrusion on their workspace. (Wardell Armstrong).

Snailbeach Mine Pumping Engine House 1993. A 60in Cornish Pumping engine was installed in this house in the the 1850s and operated until 1911; the beam was about 35ft long, stroke at cylinder 10ft and at the pump 9ft. The engine is said to have pumped a maximum of 5,000 gal/hour. By 1990 the engine house was much overgrown with trees and ivy and the stonework was in poor repair (see page 25). These photographs show the changes that took place during consolidation work, stripping off the vegetation and erecting the scaffolding.

Snailbeach Mine Pumping Engine House 1994. The scaffolding has been removed and clearance is about to begin on the vegetation and rubble in the adjoining boiler house and the winding house to the left. The boiler furnace flue connected with the flue from Snailbeach Smelter to the chimney on the hill, seen above to the left. Correspondence shows that the original chimney collapsed in July 1884 and that rebuilding took place in early 1885. The chimney is about 90ft high, hexagonal and built on original base. In the 1880s it took nearly two weeks to make a track up the hill from the mine for this work to be done.

Snailbeach Mine Powder House 1990. The mine had a substantial square shaped powder house with blast wall around it. The house was built in 1863 and had a very light roof. This and the fact that the inner and outer doors are not in alignment would have meant that for safety any blast would have gone upwards not outwards.

Early steam engine house and blacksmith shop 1994. Nearest the camera are the remains of a very early steam engine house, possibly c.1800, the oldest building remaining on site. In a photograph of about 1900, not shown, it appears to have a high timber frame in front similar to a pumping frame but no shaft is known in this position. Its purpose remains a mystery. Beyond the engine house is a blacksmith's shop, much rebuilt over the years and which could contain parts of a boiler house structure.

A 'Blow George' or Ventilating Fan 1967. This small ventilating fan was found on timber planks on a shaft on Lords Hill, Snailbeach. It consisted of a number of blades in a close fitting chamber and an outlet (bottom right) through which air was forced along a tube to where it was required. These devices, of wood or iron, were often powered by boys turning handles to rotate the blades. The fan shown was driven by a belt from a farm tractor or other mobile power source.

'The Waterwheel' near Wagbeach, Hope Valley 1993. Some 1,200 yards from the main Snailbeach Mine, the drainage level (just off the photograph to the right) enters the Hope Brook, a waterwheel here powered pumping rods underground until the 1850s. From the 1860s it was used to power the Cliffdale Barytes Co.'s mill constructed on this site. The water for the wheel came mainly from the Hope Brook using a system of leats. The photograph shows the collecting tank from which the water jet played on the wheel. By 1915 gas engines were also being used for power as there was often insufficient water available to turn the wheel. The barytes for the mill came from the company's own mines at Cliffdale. The company ceased trading in 1926. (K.C. Lock).

New Central Mine 1864. Sometimes called Central Snailbeach Mine. These photographs were taken from a shares prospectus and are the oldest known photographs of a mine in this orefield. The mine was worked unsuccessfully in the late 1860s and in 1871 and 1872; an output of 626 tons lead ore is recorded. The mine reached a depth of 164 yards and has a drainage level to the Hope Brook. The picture appears to show shaft sinking in progress with the spoiltips being newly formed.

New Central Mine 1864. The second photograph in a pair produced to gain investors for the project. The aim is to impress and the bigger and neater the engine and chimney the better. The wooden planks in the foreground appear to be covering another shaft.

New Central Mine 1965. The main building with extended sloped roof over the boiler looks much the same as in the previous photograph, except that the chimney has been shortened. A site inspection shows that at some time the building has been turned around in relation to the nearby road, about ninety degrees. Did the prospectus actually show the buildings of another mine superimposed and was the house at New Central built later to match – or was it rebuilt at some later date? A tree specialist has said that the tree to the right is over 150 years old. It could be the same tree on both the 1864 and 1965 photographs. (M. Newton)

New Central Boiler 1994. The boiler is now part of the house that has been converted to a dwelling. It appears to have been a Lancashire boiler about 38ft long and 7ft diameter. (Mr D. Young).

Three

The Snailbeach
District Railways
the tip operations, smelter and quarries

The opening in 1861 of the Minsterley Branch Line, a standard gauge railway 6.5 miles long from near Hanwood to Minsterley, provided the impetus for the development of a narrow gauge railway to the mines. The Branch Line itself was not of much help to the mines on the remote hillsides and the prospects for further extensions were not good. To overcome these difficulties two railways were authorised each of 2ft 4in gauge; No.1 from near Pontesbury Station to Crows Nest, just over three miles long, and No.2 from Crows Nest to Pennerley, nearly two miles in length. Only No.1 railway was built and this was completed in 1877. At the Shrewsbury Smithfield Sale, 4 September 1877, one lot was, '10 carthorses, powerful, upstanding, seasoned, of good ages, 17 hands high etc., for sale in consequence of the completion of the SDR which will render haulage by road unnecessary'. The Directors of the company included the Lovetts of the Snailbeach Co. and the Engineer was Henry Dennis of Ruabon. The Dennis family took over almost complete control of the railway in later years.

The narrow gauge necessitated a transhipment sidings at Pontesbury Junction and the loaded narrow gauge wagons ran over the top of standard gauge wagons on a siding beneath. From Pontesbury Junction the narrow gauge railway had to climb almost the whole way to Snailbeach. About 3.25 miles away the line ended abruptly just beyond Snailbeach village at Crows Nest.

At Crows Nest a reverse shunt provided entry to a branch nearly 1,000 yards long which included a steep slope to the mine workings. Another reverse shunt and a further 330 yards, partly up a steep incline, reached the Snailbeach Pumping Engine House. The wagons of coal for the boilers were winched up this final incline. There were also several short spurs within the mine site including one to the engine sheds.

At the bottom of the mine tips, near the point where the Snailbeach District Railway passes beneath the Snailbeach – Plox Green road, there was a further branch to a dressing plant at the base of the tips. About 1200 yards below this bridge another branch of about 500 yards went to the Snailbeach Smelting Works. Further still down the line a branch went to the quarries north of Eastridge Woods (Granhams Moor, near to the Minsterley – Habberley road), and, in later years a connection was made to a new quarry at Callow Hill, further down the line and less than one mile from Pontesbury Junction.

From 1878 to 1883 the railway made a profit; it was carrying about 14,000 tons, on average, each year for the mines but when the closures came in 1884 this dropped dramatically. For the next ten years receipts barely covered costs and when the smelter closed in 1895 things got worse. For the next ten years the railway made a loss although with the reopening of the quarry north of Eastridge Woods the situation did improve. With the closure of the deeper mine workings in 1911 and falling stone traffic during the First World War it worsened again.

A new management headed by Col. Stephens, a strong believer in minor railways, took over in 1923 and following the reopening of Black Tom Mine for barytes in 1922 and the Callow Hill Quarry in 1926, the situation improved once more. Stephens died in 1931 and by 1932 only the Callow Hill Quarry traffic remained. In 1946 steam hauled traffic ceased, wagons were then being lowered down from the quarry to the Junction by gravity and the empties were hauled back up by a farm-type tractor. Rail traffic from the quarry diminished as lorry transport took over and by 1960 had ceased to be used altogether.

The Locomotive Shed at Snailbeach about 1950. The shed lies between the mine buildings and the spoilheaps. Black Tom Mine headframe can just be seen in the trees above the left hand section of the shed. Loco No.3 stands outside the shed. (K.C. Lock).

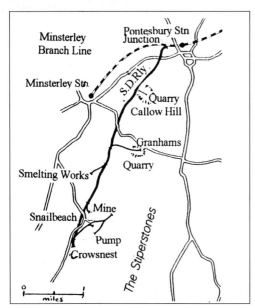

Left: The route of the Snailbeach District Railway. *Right:* Henry Dennis (1825-1906). Henry Dennis was born in Bodmin, Cornwall, but moved later to North Wales as a colliery manager. He had a number of financial interests in mines and brickworks. He became 'manager' of Snailbeach in 1870 and continued until 1881 when he became 'Chief Agent', a post he held until 1902, when it passed to William Oldfield. His interest in the SDR was taken over by his son, Henry Dyke Dennis.

54

SDR Loco Shed, Snailbeach 1949. A photograph taken three years after the line ceased to be operated by steam, and when this section of the line was no longer used. The extension to the loco shed had lost its roof as had the Mine's Crusher House behind it but the rest looked fairly intact. The company's No.3 loco stands in front of the shed awaiting scrapping. This loco was a Baldwin 4-6-0 sidetank built in America and weighed 15 tons. It was rebuilt in Britain in 1918 for the War Department and later purchased by the SDR in 1923. The loco was cut up for scrap in May 1950. The Loco shed was the first building to be repaired for use as a store as part of the 1990s land reclamation scheme on this site. Work on the building was completed in 1992 and many of the rapidly decaying small mining artifacts still on site were moved into it for safety. (Photo by Bernard Roberts, courtesy of J.A. Peden).

SNAILBEACH, MINSTERLEY,

SHREWSBURY, _____ *19*

To the Snailbeach District Railway Co.

Goods : Pontesbury Junction.

			Tons	C.	Q.	Rate.	£	s.	d.

Details from a goods ticket for the Snailbeach District Railway 1900s.

SDR between Snailbeach and Crows Nest terminus *c.*1910. At this position the line splits into two parallel tracks and the branch to Snailbeach separated from the higher of the two. The track to the right, up to the mine, was steep. The small building on the left was a small weighbridge and office and this location served as the 'station'. Officially, however, there were never any passengers. As the following photograph shows there is something wrong because the branch line appears much too steep and the bridge and the building are much too close. This, the writer is assured, is due to the camera lens playing tricks – a telescopic effect. (N.R. Taylor).

SDR between Snailbeach and Crows Nest Terminal, late 1940s. The picture was taken from a position slightly further from the Snailbeach village than the above. The camera has given a much more reasonable view with moderate inclination and background features much further away. The track was lifted shortly after this photograph was taken. (Elsie Rowson).

SDR Steam Locomotive 'Fernhill' about 1918. 'Belmont' and 'Fernhill' were the first two, of six, steam locomotives used on the line, Belmont in 1877 and Fernhill in 1881. Both locos are believed to have been scrapped before 1919. Fernhill was a 0-6-0 loco, probably named after the residence of one of the company's directors, J.H.Lovett. The reason for the gathering of young boys on the photograph is not known. (National Railway Museum).

SDR Steam Locomotive No.2 about 1935. This was one of three locos purchased second hand by Col. Stephens. It was an 0-4-2T built in 1902 by Kerr Stuart & Co. Ltd of Stoke on Trent. After a chequered history it was purchased in 1922 for the SDR. Although at the time named 'Skylark' it became simply No.2. Locos 2, 3 and 4 worked all the traffic from 1923 up to the end of steam working in 1946. Each was used, in turn, for spells of 2-3 weeks which meant, unfortunately, that eventually they all became unserviceable at about the same time. The locos were cut up by Thos. W. Ward Ltd. in May 1950. (K.C. Lock).

Reworking the Spoil Tips at Snailbeach *c*.1930. The spoil tips have been worked since about 1900 for 'spar' (calcite and quartzite chips) to be used in the building industry. From 1900 to about 1930 this was done by the Halvans Co. but other companies have worked here too, including the Gravel Trading Co. Barytes has also been extracted and milled on the tips. The photograph shows some of the equipment in use including a simple jig in the foreground, other items would have included screens and buddles. (K.C. Lock).

Reworking the Spoil Tips at Snailbeach *c*.1930. A branch off the SDR was used to give access to the railway. Here wagons are being loaded and await collection. This photograph seems to have been taken on the site near the present village hall. (K.C. Lock).

Snailbeach Smelt House or Smelter 1963. This smelter was built in 1862 and replaced the Snailbeach Co.'s smelter at Pontesford. The smelter was of the reverberatory type with five hearths. At the opening ceremony on 11 July 1862, a procession of 400 workers was followed by dinner, with ale, set out on three rows of tables 145ft long placed inside the new smelt house. At one end sat the mine purser, Mr John Jones, and at the other there was a small orchestra. The smelter closed in 1895 but one hearth was used for a short period in 1897. A tramway was used from the Snailbeach Mine about a mile away and a connection was made with the Snailbeach District Railway soon after 1877.

Part of the Smelter Flue, Snailbeach Smelt House 1961. The photographs shows the collecting or receiving chamber from the five flues in the House. From here the fumes were taken through a long flue to the high chimney on Lords Hill. Along the route the fumes passed through condensers to enable as much lead as possible to be extracted. A note in the 'engineers' book' for 2 Oct 1861 says that 518 yards of flue had been constructed with 616 yards more to go. The flue was 6ft high and contained over 500 bricks per yard. A length of flat wire winding rope can be seen on the photograph, this had been used for fencing purposes.

Minsterley. Granhams Moor Tramway. No. 2.

Granhams Moor Tramway c.1910. The connection to this quarry appears to have been rope-hauled as rope pulleys are present between the tracks on two postcards that exist, numbered No.1 and No.2. They are both similar, No.1 shows a completed length of track with rope and a junction off, No.2 (above) shows some workmen standing on a trolley in an apparently incomplete section. Both show dense woodland. (Anne Heath).

Minsterley. Granhams Moor Quarry.

Granhams Moor Quarry Plant c.1910. Granham Moor Quarry, north of Eastridge Woods, is listed in the 1896 official list of quarries although it was 'not worked' that year. It was listed as a sandstone quarry, having Percy Rowson of Snailbeach as owner. The quarry was reopened before 1905 and a fatality was recorded there in 1906. By 1917 it was a large quarry employing thirty-four persons under the Granham Moor Quarry Co. The photograph shows the scale of the quarry plant at that time. The quarry is thought to have closed after the First World War. (Anne Heath).

Callow Hill Quarry 1941. This photograph shows a mishap during loading at the Callow Hill Quarry in 1941. In 1948 the County Council was producing 'Basalt' class aggregates here and they were the only user of the shortened length of the SDR, steam trains having stopped from Snailbeach in 1946. The quarry became less dependent on railways and amounts began to reduce. By the end of the 1950s all material went by road. The quarry continued to be operated by the Council until the early 1980s but has since been operated by Tarmac Roadstone. The workers in the photograph are believed to be Charlie Hayward and Percy Crowther.

Callow Hill Quarry 1963. Callow Hill Quarry opened in 1926 and was then operated by Haywards Quarries Ltd. A two-foot gauge railway operated in the quarry itself; the wagon loads of igneous rock being pushed by hand to the crushing plant under which ran a SDR siding. The SDR wagons were then loaded directly from the plant. In 1931 the County Council leased the quarry and bought it in 1956. The photograph shows the only railway wagon seen in the Quarry in 1963 (with the author's wife 'for scale').

Transportation of aggregates from Callow Hill to Pontesbury 1954. This is a very rare photograph showing the use of gravity for taking down the loaded wagons, the brakeman unwisely riding between them. After emptying them at the Pontesbury Junction transhipment point, the empty wagons would be returned eventually, in batches, drawn back up the track to the Quarry loading area by tractor. The rider is believed to be J.A. Rowson. (A. Pratt).

Tractor and train in the 1950s. The railway's tractor is hauling empties over the Shrewsbury Road , which isolated the main railway from the transhipment point. The bridge, which has the date 1875 has now been removed and the SDR line converted into a quarry access road. The view under the bridge is towards Snailbeach. The workers are believed to be Mr J. Rowson and Mr Edwards. (Emily Griffiths Collection).

SDR Wagons over transhipment sidings, Pontesbury 1962. These bottom-emptying wagons were standing over the vacant standard gauge siding, the rails supported on heavy timbers. The wagons were later removed for preservation.

Pontesbury Junction Transhipment Point 1962. This photograph shows a rather empty site, except for the tractor which was used for hauling the wagons back to the quarry, its wheels astride one of the rails. At the time of the photograph the rail track was being lifted.

A steam train in earlier days (1940s). Empty SDR wagons are being taken back to the mine over the Shrewsbury road by one of the Baldwin steam locomotives.

Consignment note from major user, Pontesbury Junction 1900s. A ticket with 'Halvans Company Ltd.' heading; for 'ores' to be collected in railway wagons from the transhipment point of the SDR.

Quarry Workers at 'Pontesbury' Quarry c.1910. It is not definitely known at which quarry these workers were employed as there were several near Pontesbury, the two most likely were however Granhams Moor or Nills Hill. Callow Hill was not open at this early date. Several workers are holding quarrying tools. Most look clean and well dressed. Nills Hill Quarry, was operated by the County Council and in 1896 employed forty-two persons, it was still being listed as a quarry in the 1960s directories but closed soon after.

Four

Tankerville, Pennerley & The Bog Mining Areas

These mines are all, like Snailbeach Mine itself, on the eastern outcrop of the mineralised Mytton Beds. By far the most important mines were the three named but there were others opened with great enthusiasm because of their proximity to the lines of the veins worked in these three successful mines. The principal workings were all served by an underground drainage level about two miles long, the Boat Level.

Tankerville Mine is on the site of an older eighteenth century mine called the Ovenpipe Mine and its location may well have been influenced by the promoters of the Boat Level, a drainage and navigation level. The Ovenpipe Mine could not have been large or important as it is not mentioned on any pre-mid-nineteenth century map and is thought to have been a small pipe-working. The driving of the Boat Level, about 371ft below the surface at Tankerville, may have provided sufficient reasons for the development of the larger mine.

The Boat Level continued on to the more established Pennerley and Bog Mines, two of the earliest in the area. For most of their history these mines were worked together. In later years the Boat Level was taken beyond The Bog Mine but no really successful operations were carried out there.

On the surface the three mines are connected to Snailbeach by a narrow, winding road along a shelf on the flanks of the Stiperstones range. There were small communities at Bog, Pennerley and Tankerville but Snailbeach was certainly the most important and very little remains at Pennerley or The Bog today. A group of small mines developed around each of the larger mines but none met with great success.

Tankerville Mine c.1955, Ovenpipe and Tankerville Chimneys. The mine was originally the smaller Ovenpipe Mine worked by the Stiperstones Co. but from the 1860s it was worked with adjoining mines such as Potters Pit and Pennerley Mine under the name Tankerville Great Consols. The most active period was between 1865 and 1891 when 17,924 tons of lead ore and 3,049 tons of zinc were produced but this may also include some output from the smaller mines. In addition the mine also produced some barytes, witherite and silver (from the lead ore). Recorded employment reached a peak of 171 in 1879, of these 112 were underground.

Tankerville Mine about 1870. At this time the Cornish Pumping House had not been built and this and the Watsons Shaft were not completed until 1876. Ovenpipe Shaft is to the right and the workmen are standing in front of the new block of buildings for Watsons Shaft. It is interesting to see that so many of the early buildings were of wood. The headframe arrangement at Ovenpipe Shaft (to the right of the photograph) is unusual in that the pulleys carrying the ropes from two winding engines are almost at right angles to each other. This is the reason why the two winding house chimneys appear close to each other on the following photograph. (K.C. Lock).

Tankerville Mine about 1870. Probably taken at the same time as the above, this shows a different view with the mine reservoir in the foreground and the mine yard at centre. To the left of the wall, on the left of the photograph, there was (and still is) a fine row of orebins and beyond them, just off the picture, the site of the Cornish Engine which was not built for six or more years. (K.C. Lock).

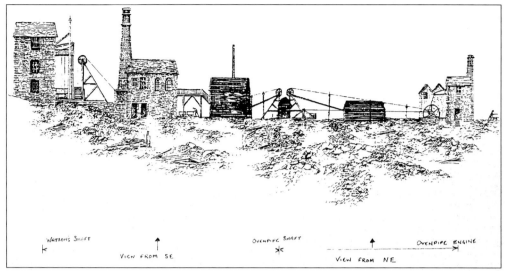

WATSONS SHAFT   ↑   OVENPIPE SHAFT   ↑   OVENPIPE ENGINE

VIEW FROM SE   ✕←   VIEW FROM NE

Tankerville Mine 1893. A sketch based on a drawing on the abandonment plan. Compare this with the previous photograph; the Cornish Engine house has now been built on the left and the chimney, stone winding house for Watson's Shaft, the wooden winding engine house and headframe in Ovenpipe Shaft remain. Ovenpipe winding engine's stone house and chimney are off to the right on the sketch but this was a licence on the part of the original artist - this line of buildings is in fact almost at right angles (note the position of the small chimney at Ovenpipe on the previous photo). (M. Newton).

Tankerville Mine Buildings about 1963. The same line of buildings as at the top of previous page but now with the 1876 Cornish Pumping House in place but in ruins, its chimney to its right. The remains of Watson's shaft stone winding house and the attached building to its left can be seen. Ovenpipe wooden winding house has gone but Ovenpipe Shaft's remains on the extreme right are marked by a fence and large bush. (K.C. Lock).

Tankerville Mine 1997. The Cornish Engine house about to be consolidated and the chimney complete with scaffolding. The shaft here became the deepest in the orefield at 1,612ft below surface. The vein structure was unusual as it formed a 'vertical' pipe. There was a steam engine underground at 1,140ft giving what was probably the longest 'chimney' in the country albeit a part of the shaft. The mine finally closed in 1893 and the equipment was sold by auction in 1902.

Tankerville Mine Cottages 1893. Sketched from the drawings on the abandonment plan. From the right are New Shaft, Watsons Shaft (with Cornish pump) and, centre, Ovenpipe Shaft. Very few dwellings were provided for workmen by employers in this area so those shown on the left are unusual. The lower drawing is a detail of the cottages, the first three dwellings are still in use, the middle of the row is demolished but the buildings to the left remain. (M. Newton).

Tankerville Mine area 1997. A rare view taken from the top of the chimney showing particularly well the chapel-like Count House or Agent's Office which it is said had a large upper room for use as a Club Room for the miners and for occasional religious services. In front of the Count House is Ovenpipe Shaft, marked by a group of trees, to the left of the photograph is the end of Tankerville cottages – a row of houses built for mine staff of various levels.(see drawing on previous page). Behind the row was a belltower and a 'flagstaff' stood on the mound. There is another shaft 'Lewis' behind the cottages at the top of the photograph.

Arthur Waters 1825-1887, Mine Agent. Born in Goldsithney, Cornwall, Arthur Waters came to Shropshire to work for the Stiperstones Co. when aged twenty. He worked at Pennerley, Ovenpipe then Roman Gravels Mines (where he was agent) – in fact, at some time, he worked at seventeen of the mines in the area. Capt. Waters played a major part in the development of Tankerville; he was a hard worker and much respected. His death at the age of sixty-two is said to have been due to his efforts in trying to keep the mines going in difficult times. In his later years he worked with his son, also Arthur, as A. Waters & Son, and his son continued as a mine Capt. after his father's death. Arthur Waters (Snr) was buried at St Hilary, Cornwall, his home county. (K.C. Lock).

Burgam Mine 1959. Burgam Mine is situated close to Tankerville; it was always a small mine and probably began in the early nineteenth century. It consists of several shafts, adits and has its own drainage level. The Boat Level lies about 90ft below this drainage level providing the mine with extra drainage. Periods of activity are known from the 1860s and the peak was probably between 1890-1894 when up to eighteen men produced 2,026 tons of barytes, 21 tons of lead ore and 4 tons zinc ore. Intermittent working took place during the period 1920-1945 when ore and barytes was taken down the hillside to the road by horse sleigh. Some new levels were driven here for exploration purposes between 1957-1962, the last real mining carried out in this district. The photograph shows the mine manager, Tom Rowson, drilling a shothole using a manual machine.

Burgam Mine 1960. This old mine is shown during its last working by J.L. Burden, under the name Jackfield Mines Ltd. A number of short adits were driven into the hillside by two local miners, Tom Rowson (manager) and Norman Evans, with occasional assistance from others. The equipment was primitive, a wooden shed for office and cabin, and a tin bath was used for washing ore and the miners. The tip and two mounds of ore and spoil can be seen on the photograph. *Inset: Burgam Mine, Horse and Sleigh 1940s.* Burgam was one of the hillside mines where a horse and sleigh (called locally 'a car, a strong wooden box on two steel runners') was used to bring the ore down the hillside to a loading point for trucks on rails or lorries. This is a sketch by Clifford Lewis, a former Burgam miner, as he remembered the mine before it closed in 1943. Jack Evans is leading his horse, also named Jack. Mr Lewis left the mines and some years later, when homesick in Australia, drew this sketch, and many more to remind himself of happy times back home.

Burgam Mine 1959. The mine worked mainly the Big Spar Lode in which the chief mineral was barytes. The Manager, Tom Rowson gleefully holds specimens of lead ore at one of the adit entrances.

Burgam Mine 1959. The author of this book was an occasional helper during his spare time from his employment at a Shropshire coal mine. Timber cut locally was used for support and candles were kept burning at intervals along the adits for lighting. There was no machinery, no safety wear and no real urgency, a great contrast to work in a colliery. Working at the mine became intermittent and then stopped in 1962 due to 'financial problems'.

Roundhill Mine about 1907. Roundhill Mine was at work through much of the period 1846–1868 as a lead ore producer and then in the 1890s as a small barytes operation. In 1905 the Earl of Tankerville reopened the mine employing up to twenty-five men for exploration and production partly as a means of finding work for the unemployed. The mine worked for short periods after this, producing barytes, until about the time of the First World War. The dumps were again worked for calcite in 1924 and 1940. The steam engine used here is said to have been built at Perran Foundry in Falmouth in 1874. (K.C. Lock).

Roundhill Mine c.1920. These workers are believed to have been erecting the aerial ropeway as it passes near Roundhill Mine (note the ropeway support to the left of them). This ropeway was a feeder branch from the Buxton Quarries to the main Bog – Minsterley ropeway. The miners were: back row, R. Rowson; J.G. William; G. Williams; -?-; T. Hill. Front row -?-; S. Purslow; J. Purslow; R. Purslow; T.Lewis. (SCMC).

Please bring this Catalogue with you.

AT PENNERLEY
AND
TANKERVILLE MINES.

About five miles from Minsterley Railway Station, half-an-hour's run by rail from Shrewsbury.

CATALOGUE
OF THE VALUABLE

Mining Plant and Machinery.

MR. J. E. DAVIES

is instructed to SELL BY AUCTION at the Mines above-mentioned, On FRIDAY, September 5th, 1902, at 12.30 o'clock, the whole of the

Valuable MINING PLANT & MACHINERY,

including:

Two Cornish PUMPING ENGINES,
60in. and 40in. Cylinders, by *Harvey*, Cornwall.

Massive BALANCE BEAM,
with extra strong Shafting and Pedestals, by the same maker.

Six Horizontal and Beam WINDING ENGINES,
varying from 9in. to 22in. Cylinders, by eminent makers.
Winding Drums, fitted with Spur and Cog Wheels; Powerful Capstan Gear.

Six Double and Single Tube GALLOWAY and CORNISH BOILERS.
12in. Air Cylinder on Iron Bed Plate.

Valuable CRUSHING PLANT,
with the usual Shafting, Strap Pulleys, Rolls, Sieves, &c., &c.

Valuable DRAWING LIFTS, 15in. Diameter,
Plunger Poles; H Pieces; Working Barrels and Winch Bores; Ditto, do. 6in. and 7in. Pumptrees.

PITCH PINE HEAD GEARS, with Winding Pulleys.
Several Coils of Wire Rope, in varying lengths; Several Tons of Bridge Rails; Flange Steam and Feed Piping of varying dimensions.

DRESSING PLANT,
comprising Circular Buddles with iron centres; Patent Jigs, with Shafting and Strap Pulleys; Patent Saw Bench, 5ft. 6in. by 2ft. 6in., with four Circular Saws.

Double Purchase CRAB WINCH.
Three Patent Rock Drills; about 12 Wrought Iron Kibbles, several Pit Tubs; Corrugated Iron and Wood Erections; many tons of Wrought and Cast Scrap.

The Contents of
Fitting, Blacksmith's, and Carpenter's Shops, &c., &c.,

These Catalogues may be obtained from Messrs. How and Son, 9, Swan Hill, Shrewsbury; or from the Auctioneer, 11, Wrexham Street, Mold.

Conveyances will meet the Trains at Minsterley Station to take intending Purchasers to and from the Sale at reasonable charges.

W. M. BELLAMY AND CO., LTD., PRINTERS, MOLD.

Front page of Auction Catalogue, Pennerley & Tankerville Mines 1902. Pennerley mine is a very old mine, records going back over 200 years. The Mine has been worked very much on a stop-go basis, often in association with Bog Mine which has tended to overshadow it. It has, for short periods, been a very rich mine producing good quantities of lead and zinc (1870s-1890s) and barytes (from 1890). The peak year for employment appears to have been 1883 when 146 persons were at work. Pennerley mine was over 1,080ft deep in 1868. In the 1870s the mine was worked in conjunction with Potters Pit by the Stiperstones Mines Co. It had four principal steam engines, a 60in pumping engine, a horizontal engine for lifting pump rods and repair work, a third engine for winding and crushing and a fourth engine for winding at the other three shafts – one of which was Potters Pit over 300 yards away. This company went into liquidation by 1878 and the mine was taken over by Tankerville Great Consols, who in 1880/81, put in an air compressor and an aerial ropeway to bring ore from Potters Pit. Falling lead prices however forced closure on the mine and the equipment was auctioned in 1902.

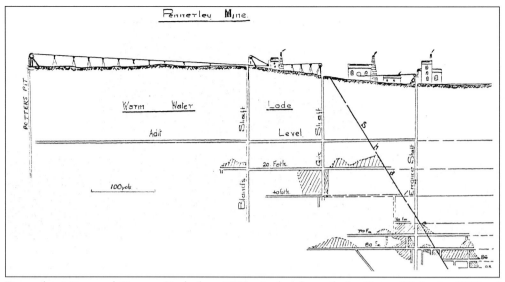

Pennerley Mine and Potters Pit before 1875. A sketch made by Percy Blight in 1920, of drawings and locations of the buildings as shown on the mine plan. The depth to the levels is given in 'fathoms', the position of the drainage adit (Boat Level) and the workings in the Warm Water Lode are also shown. (SCMC).

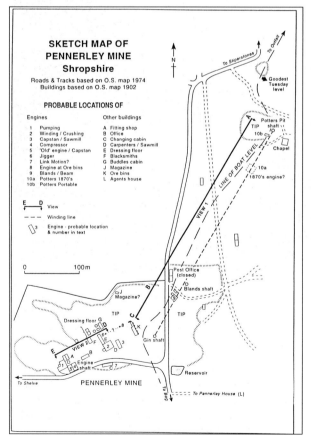

Sketch map showing location of surviving remains, 2000. This is based on OS 1902 Map and a paper in Bulletin of the Peak District Mines Historical Society Vol.12 No.2 1993. In the year 2000 the surface area of the mine appeared as derelict land, fairly level and much scraped over, with a few clumps of trees. In fact, each clump of trees on the main site represents one of its buildings, the excavators having worked around the substantial foundations of each building. This would be an ideal site for archaeological investigations, not only do the foundations remain but the history of each building on the site is so well documented.

Pennerley Mine – the surviving building 2000. This building stood remote from the others and survives on its own heap of rubble. Due to its remoteness and survival it has been called by many the *Powder Magazine*. This may be so – but unlike other 'magazines' it is not recorded as such on OS Maps and does not appear rugged enough to store explosives. Furthermore it appears to contain ironwork in its structure and to have had a window – both unwise and unusual in magazines. It could well have been an office or store.

Potters Pit Engine House and Miner's Cottage *c.*1900. Potters Pit was probably opened in the eighteenth century and was a pipe-working but much smaller than the Tankerville Pipe. It is said to have been named because its ore was especially suitable for use by 'potters'. Like Pennerley it is connected to the Boat Level (at about 330ft deep). The lowest workings were described as being very hot but not particularly deep (about 960ft). The mine worked in its own right between 1857 and 1868 and, at other times, intermittently with the Bog or Pennerley. It was connected to Pennerley Mine, 350 yards away, by a winding rope, an aerial ropeway and a navigation level at various times but the engine house shown on the photograph remains from a later period. It housed a 'portable' winding engine. The 'thatched' cottage had its roof raised and tiled early this century and has recently been almost completely rebuilt. It lies close to the still open Pennerley Chapel and the Crown Public House (in ruins). (A. Tucker).

The Bog Mine c.1910. This is a very old mining site and The Bog Mine is recorded as early as 1739. By 1777 the Mine had a Boulton & Watt steam pumping engine with a 30in diameter cylinder. The mine had a chequered history and by the end of the nineteenth century employed only nine men producing lead and zinc ores and barytes. From 1907 operations expanded considerably under Bog Mines Ltd, then the Shropshire Mines Ltd, employing at first about seventy workers, then over 100 by 1917. A new shaft was sunk, called Ramsden shaft, and the first electric winder to be installed at any Shropshire mine was placed here – although it was unsuccessful and replaced by a steam engine. The photograph shows a tall new building in the centre and this contained 'National' gas engines driving Crompton Parkinson alternators to produce electricity, the gas for the engines was made on site from coal and anthracite brought from Wales. Electricity was also used to light the village – the first village to have such services in the County.

Much of the work in this period was done around Bunting Shaft (the headframe can be seen behind the buildings to the right) and the principal mineral produced was barytes. Official records show that between 1910 and 1924 there were often as many men working on the surface, in the dressing plant and new developments, as there were underground. About 1917 an aerial ropeway 5.5 miles long, was constructed between The Bog and Minsterley using German Prisoners of War as labourers. The mine closed in 1924 but some work did continue at the nearby Rock Mine.

Mine and Village in 1962. Compare this photograph taken from the carpark of the now closed Miners Arms, with the picture above. The Barracks or Hostel for workmen, with the long sloping roof survives (right). The school and cottages (centre) and the The Bog Men's Club buildings are shown. Only the school building survives in 2000, as a Field Centre.

Bog Mines Platoon, C Company, Kings Shropshire Light Infantry *c.*1916. The manager and his men formed a 'volunteer' platoon in case they were needed. They were photographed in front of the mine buildings as are shown in the picture on the opposite page. The mine produced between 2,000 and 8,000 tons barytes each year throughout the war. The poorest year was 1917 when it fell from 7,046 tons in 1916 to 2,315 tons. The long-roofed barrack house and power house are clearly seen behind the platoon.

Bunting Shaft Headframe 1960. The shaft has been used for both pumping and winding. It is connected to the Boat (or Drainage) Level at 350ft depth. It is believed that this wooden headframe was constructed in 1910; shortly after the photograph was taken the headframe was dismantled and the area partially cleared.

Sales by Auction.

To Mining Adventurers are offered the valu-
able LEASE of the BOG LEAD
MINES, Steam Engine, Gins, Boats,
&c. the Property of Mr. John Lawrence,
jun. a Bankrupt.

BY MR. SMITH,

On Wednesday, the 24th of February, 1830, precisely
at Twelve o'Clock, on the Premises, at the Bog
Mine, in the County of Salop ;

LOT I.

THE valuable LEASE of those inex-
haustible LEAD MINES, called the BOG,
extending over a Mining District of upwards of Three
Thousand Acres.—The Work is open, and there is a
navigable Level for Boats that Drains at One Hundred
and Fifteen Yards. And there remains Two Thousand
Four Hundred Tons of Ore to be raised, free from
Royalty.

LOT II.

In the following or such other Lots as may be agreed
upon at the Time of Sale ;

A capital STEAM ENGINE, 42 Inch Cylinder
(Double Power), with 2 Wrought Iron Boilers, Steam
Pipes, and 55 Yards of 15-Inch Pumps, Working
Barrel 14 Inch, with Plates for Pump Rods and Joints,
Part of 2 Winches, Pair of large Pit Blocks, Cap-
stan Frame, valuable Capstan Rope about 200 Yards,
Timber Hanging Rods with Iron Work for the Engine
Pumps, Cast Iron Rails and 2 Rail-way Waggons
under Ground, 3 Gins, Ropes, &c. 3 Wood Boats and
1 Iron Boat, with various other Articles.

For further Particulars apply to Mr J. W. WATSON,
Attorney; Mr WM. HAZLEDINE; or the AUCTIONEER,
all of Shrewsbury, if by Letter, Post-paid

The Bog Mine, Auction of Lease 1830. This newspaper cutting shows the lease of the Bog Mine to included a 'Navigable level for Boats' and 'two Thousand Four Hundred Tons of Ore ready to be raised'. Lot 11 gives a list of articles, this includes '3 Wood Boats and 1 Iron Boat' which seems to remove all arguments that the 'Boat' Level was never actually used as such.

Boat Level Explorers 1967. The Boat Level (or Drainage Adit) commences in the valley below Tankerville Mine at a altitude of 892ft OD. It was driven from there, starting about 1790, then under the mines at Tankerville, Pennerley and the Bog. In the early years of the twentieth century it was continued from the Bog about 450 yards towards Ramsdens Shaft and proposals were put forward to go further. Total length was about two miles.

A section of the Boat Level from its portal was opened up in the 1960s and explored as far as Tankerville Mine where there was a serious blockage with water shooting out in fountains. The rather wet explorers shown in the photograph are: the author; I.J. Brown; -?-; M. Summerfield; D.R. Adams – all members of the Shropshire Mining Club. Another 'wet' explorer that day (Ken Lock) does not appear, he was behind the camera!

The Men's Club 1983. The Bog Men's Club was constructed in 'timber and sheet' early this century and the ground floor was used as mine offices. The building was founded on the remains of old stone buildings which could have been a boiler house. It had two floors, two rooms on each with space for mine offices and two billiard tables, table tennis and dancing. It closed as the population fell and the contents were sold by auction in 1975 (see inset.) Soon afterwards it was demolished except for the stone walls, which now hold a number of interpretative panels giving information on the social and natural history of the area.

Ritton Castle Mine engine house remains 1963. Close to The Bog Mine in a secluded valley lie the remains of a mid-nineteenth century promotion put forward with great optimism but which actually produced little. Great store was put on the unsubstantiated view that all the good veins of the successful mines converged on this point. The engine house and chimney were built about 1855, for a Harveys of Hayle engine, and prospecting continued for about a decade. These remains and a small earth dam, used to form a reservoir for a waterwheel in the dressing area, are all that remained in the 1960s.

Ladywell Mine, 1963. Although earlier shafts had been sunk here without much success a major attempt at working was made in 1871. By 1874 the new engine shaft had been sunk. In 1875 a steam engine house was erected and an engine installed capable of both pumping and winding. It was a traditional beam engine with crank and flywheel to give rotary motion. Some ore was found but not enough even to pay dividends to shareholders and in 1882 the company was liquidated.

The photograph shows how the house looked in 1963, it has recently been repaired and consolidated, note the long slit for the flywheel. A similar arrangement existed at East Grit Engine House.

80

Five

Ladywell, Roman Gravels and the Grit Mining Areas

It is a widely held belief that the mines of this western outcrop are in the oldest worked area of the orefield particularly those around the Gravels and the Grit. There is, however, little documentary information to support this yet, much of it is based on the lead ingot and old tools that have been found.

This western area had certainly become a centre for mining by the late eighteenth century and early nineteenth century as is shown by the purchase of Boulton & Watt engines for both Gravels and Grit mines and the records of the Lawrence family and law suits. The Grit engine of 1783 (see page 12) was the second recorded Boulton & Watt Engine in the orefield and others soon followed.

The driving of the Wood Level, a drainage level driven from the floor of the Hope Valley, began under John Lawrence in about 1790. At first it passed under the Wood Mine and Roman Gravels Mine sites but was progressively extended under Ladywell and as far as Old Grit. The driving of this level, about 1.75 miles long, would have done much to explore the vein structure of this area and add to its riches.

At its deepest point, the Wood Level could only drain to about 250ft depth, so to gain more depth and more drainage an even more ambitious scheme, the Leigh Level Tunnel, was commenced further down the valley in about 1825. The Leigh Tunnel Drainage Co. proposed to drive a five-mile long adit to drain all the mines in this part of the orefield and began at Leigh Hall, 1.5 miles from the mineralised ground. This proposal was fought by John Lawrence, of the Wood Level, through several law suits. Lawrence won but the cost was so great that he was obliged to sell all his lead mining interests. The Leigh Tunnel Drainage Company had also used up much of its capital in these lawsuits and had to suspend activities at about 1.25 miles and before it reached the mineralised ground at the Batholes Mine.

About 1918 Shropshire Mines Ltd, decided to continue the tunnel and in 1919 sank a new shaft, Milnes Shaft, 14ft diameter, near Batholes Mine and to tunnel from there towards Wood Mine and East Roman Gravels Mine. When it reached the Wood Mine, the Leigh Tunnel would be 3,932 yards long and a further 1.5 miles would have completed it to Grit Mine. The tunnel was driven using the most modern methods of the time including compressed air drilling, blasting and battery locomotive haulage. Unfortunately in 1923, at about two miles length, work ceased near Wood Mine. The drainage advantage to the workings is obvious when it is realised that the Wood Winding Shaft surface level here was 987ft OD and old Wood Level drained to about 100ft depth. The Leigh Tunnel would be at nearly 600ft depth (500ft difference) but this was only about half the depth of the former workings in this area even at that time.

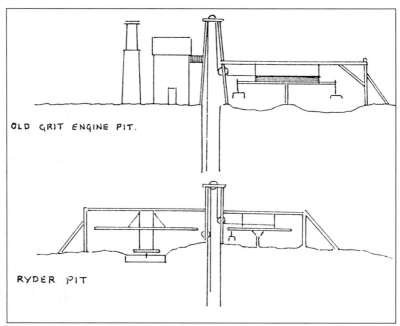

OLD GRIT ENGINE PIT.

RYDER PIT

Horse Gins on Grit Mines 1890. There must have been many horsegins in the area but none seem to have been photographed although some drawings are shown on old mine plans and sections. These two gins were originally traced from such a plan in 1965, the upper one appears to have been used for raising and lowering materials at a pumping shaft (a manually operated capstan was often used for this task) and the lower for winding only. Horsegins are also shown on mine sections dated 1844 of Wood Mine, at Whim Shaft (Whim was a type of gin) and Engine Shaft. A 'Round House' is shown on early maps of Black Tom Mine, which may have been an enclosed horse gin such as was found at some large farms. (T.J. Davies).

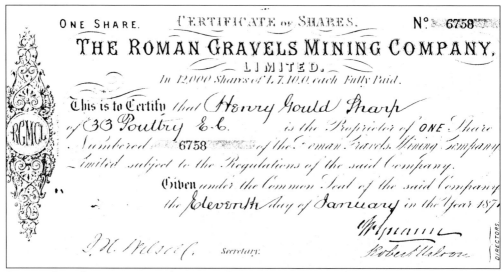

Roman Gravels Mining Co. Share Certificate 1871. Several share certificates of this time survive; they were issued to raise additional capital for this rather successful company. By 1881 the mine reached a peak employing 321 persons producing nearly 3,000 tons of lead ore (with silver), zinc and barytes in addition.

The Grit or (or Gritt) Mines – White Grit Engine House 1964. There are three principal mines at the Grit, all with surviving engine house remains, White, East and Old. The White Grit Engine House is the best preserved engine house of its era in the orefield although it may have been adapted for later engines. Some experts believe that it may also be 'the oldest surviving Boulton & Watt engine house in the world'. The earliest engine was installed at this mine in 1783 and is believed to have been replaced in the 1840s. The mine worked intermittently until about 1905 and the shaft was over 600ft deep.

East Grit Engine House 1970s. The East Grit (or East White Grit) Mine seems to have been expanded considerably in the 1850s but was normally worked in conjunction with White Grit. The engine house is likely to date from about 1860, it is large, with a large shaft, and, having the slit down the wall, is similar in appearance to Ladywell Mine's engine house. (M. Newton).

Romans Gravels Mine about 1890. This mine was called Shelfield Gravels or just 'Gravels' mine until the mid-nineteenth century but with the finding of some old tools and pottery and general interest at the time in the Romans, who were reputed to have worked here, the name 'Roman Gravels' was adopted. The mine is certainly of great antiquity and proved to be rich in ore even in the late nineteenth century. By 1890 the mine employed 200 men, fifty more than Snailbeach at that time.

The photograph shows the three-decked bridge at Roman Gravels with upwards of 100 men standing on the top deck. The waste and ore passed over this bridge to the extensive dressing floors on the other side of the road. To the right of the picture is the Old Engine Shaft headframe, this shaft being used for both pumping and winding ore. The buildings in the foreground were the carpenters' and smiths' shops. The mine closed about 1910.

Roman Gravels Mine about 1919. This photograph shows the scale of waste tipping at the mine, much of which has, in more recent years, been removed as hardcore. Roman Gravels Mine Old Engine Shaft and buildings are to the right, mid photograph. From them continues Murchison's Way up to the quarries and huts on the hillside. Above these is the New Engine Shaft with its Cornish pumping engine house and its ancillary buildings and tips. The bridge, and many of the lower buildings, were blown up in explosives trials during the First World War and some of the remainder during the Second World War.

Roman Gravels Mine 1967. The three-decked bridge crossed the road immediately to the left of this photograph, the narrow road to Bentlawn, passing Roman Boundary Mine can be seen crossing the tips. In the foreground are the remains of the pumping and winding engine houses at Old Shaft with boiler house foundations and other mine structures including two ore hoppers. Visitors to the mine in the 1880s reported several waterwheels (two of 30ft diameter), several steam engines and a 'device like a flattened cone over the centre of which the water used for washing the ore was passed'. This separated out the slime-ore which 'now helps in no small degree to swell the dividends of the proprietors, whereas formerly, this valuable source of profit flowed away, poisoned the brooks and killed the fish for miles'. Much of the tip area has now been restored to poor grazing land.

Roman Gravels Pumping Shaft 1967. A rare picture of the large wooden pumping rods still in place in the New Engine Shaft; it appears as two rods because part had already broken off, these collapsed down the shaft shortly after the photograph was taken. The shaft was sunk about 1875 and a large 60in diameter cylinder Cornish Pump, by Harveys of Hayle, was installed. It is believed that all pumping had ceased by 1900.

East Roman Gravels Mine 1975. This mine, the next sett down the valley from Roman Gravels, has been variously called, Upper Batholes Mine, West Tankerville, Hope Valley Mine and Wood Mine. While Roman Gravels Mine was declining East Roman Gravels was rising and, in 1898, reached its peak of employment at over 100 men producing lead and zinc ore. It was not a new mine but had been worked extensively before, particularly in the 1870s. This last push was shortlived, however, and nearly all underground working had finished by 1900 although processing of tips continued for some years. The photograph shows a side view of the pumping engine house with workshops on the right, all built in local stone. (K.C. Lock).

East Roman Gravels Mine 1975. The pumping engine house at Wood Shaft with part of the boiler house to the right. This was an old engine, possibly early 1850s. The Wood Level was used to reduce pumping costs at this mine, and in the 1920s there was an attempt to drain the mine to even greater depths from the Leigh Level. The Wood Level was at 124ft depth and the total mine depth was about 500ft. (K.C. Lock.).

East Roman Gravels (Hope Valley) Mine *c.*1905. This was part of East Gravels Mine, further down the Hope Valley (north) and the old postcard shows the Wood Winding Shaft with headframe and the dressing plant for the mine. The number of workers fell rapidly after 1900 and, from 1905 to the First World War, less than a dozen men were employed. The plant and the tips are most extensive but no details of the plant have yet been found. The cottages on the roadside still survive although considerably altered.

Hope Valley Mine 1980. The same view as the above taken about seventy-five years later.

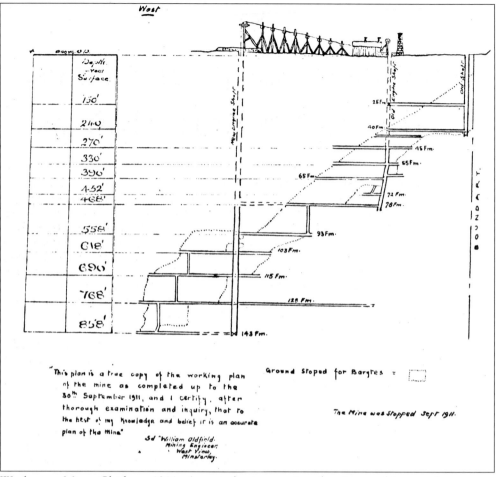

Wotherton Mine, Chirbury 1911. A typical mine section showing workings as drawn on a mine's abandonment plan in 1911 and signed by the manager. This is a tracing produced by Joseph Blight, Surveyor for the Shropshire Mines Ltd, 17 December 1919. There were two principal shafts. New Engine Shaft was some distance away with rope or chain supported on 'rolley posts'.

Wotherton was the largest barytes mine in the area up to 1910, employing over sixty workers and producing as a maximum 6,100 tons in 1910. The date of opening is not known, but before 1839, when the deposit is mentioned by Murchison. In 1865 the Wotherton mine is recorded as having produced over 1,000 tons of barytes and later two other smaller Wotherton Mines are mentioned in the records as well. The principal mine was operated by the Wotherton Lead & Barytes Mining Co. but despite the name it is believed that no lead ore was ever worked there. The mine closed in 1911 except for some surface working. The Barytes here was so pure that it needed only handsorting and grinding at the mine before being transferred to the mill at Hanwood.

Six
Barytes & Copper Mines, Mills and Smelt Houses

Although barytes is mostly found on the eastern and western flanks of the lead and zinc producing area some barytes was actually found and worked at the lead mines. As a mineral barytes was not in great demand during the peak lead mining years in the County but its production did serve to help the mines during the lean years. Normally less than 2,000 tons was produced per year until 1873 but up to 5,000 tons per year was produced up to the end of the century. From this date it increased rapidly to nearly 14,000 tons in 1913. In the 1930s Huglith Barytes Mine alone was producing up to 20,000 tons per annum with very little from other mines. During this period of production Shropshire produced between a quarter and a third of Britain's total production annually.

The mineral was usually taken to a mill for crushing and purifying and then was used as filler for earthenware, paper, paint and many other products. The mineral is naturally white to pink and sometimes bitumen staining is found.

Occasionally the mills have been built at the mines but the bulk of the output has gone to specially contructed central mills as at Cliffdale and Hanwood and, from about 1922, Malehurst. A variety of modes of transport has been used; road (using horse, steam tractor haulage and later lorries), by narrow gauge railway (the Snailbeach District Railway and the Cothercott Railway) and by overhead ropeway, (Bog to Minsterley, with at least two feeders, and the Huglith to Malehurst ropeway).

At the lead mines barytes had to be separated out, as did other useful minerals such as witherite, fluorite and calcite. At some barytes mines small amounts of lead ore were also separated and sold.

Copper was sometimes found with barytes and at one mine, Westcott, it was produced, but less than 100 tons of ore per year, from 1866-1868. The other copper mines; Wilderley, Norbury and Chittol Wood were little more than 'prospects'.

Only the ores of lead have been smelted within the area; this must have been done from the earliest lead mining times but it is first described by Hoosen in 1747. A smelt house was set up at Malehurst in the late eighteenth century and replaced by one near the Horseshoes Inn, Pontesbury (Lawrence's) before 1831. Another smelt house was set up at Pontesford by the Snailbeach Co. shortly after 1784 and was later replaced by one at the mine. A further smelt house at Pontesford was erected by the White Grit Co. before 1842 but had closed by the middle of the century.

Wotherton Mine c.1895, Showing the heaps of barytes between the road and the mine. (K.C. Lock).

Wotherton Mine *c.*1900. A group of miners at the mine. Nearly all the miners wear some kind of muffler and a hard hat with 'gob of clay 'supporting a candle. (Shropshire Records & Research Centre).

Wotherton Mine *c.*1910. A further group of miners in front of the mine buildings; there are no hard hats, some mufflers and many moustaches. There is a dog and a young boy in the front row but they do not seem to be connected with the same man – perhaps they had come to 'fetch' the men home. The dog is possibly common to both this and the photograph above. (T. Gwilliam).

Wotherton Mine (?) c.1900. A photograph said to have been taken at the mine c.1890. Some of the boys in the back row appear rather young but many seem to be in their 'Sunday Best'. The hats are unusual for miners too. One man to the right is holding a large barrel probably used for milled barytes. There is another barrel

behind the men on the extreme right. (Shropshire Records & Research Centre).

Hanwood Barytes Mill c.1920. The white barytes from practically all the mines to the west of the Stiperstones except for Weston and Calcot Mines was dealt with at this mill. There was a corn mill on the site before 1800 and by 1810 it had become a Flax, Yarn and Thread Mill. By 1886 it had become a barytes mill owned by the Wotherton Barytes & Lead Mining Co. In 1915 the processing consisted of crushing, bleaching (using sulphuric acid), washing and milling using a pair of rolls. In 1919 the mill was operated by The Shropshire Mines Ltd whose stated policy was 'to produce the best quality of ground barytes which science can devise'. The mill closed in 1922. In the foreground are the tramways used to take materials from the nearest road to the mill.

Other barytes mills included those at Cliffdale, Sutton Mill (Shrewsbury) and Malehurst. The latter worked from 1922 to 1948 and when this opened all the other mills closed. Since 1956 Malehurst Mill has been used for the preparation of animal foodstuffs and for other industrial uses.

Wotherton Mine, Chirbury, 1967. This dwelling has been converted from an old steam engine house, the rounded groove made for the flywheel is quite clear. It is not shown on the section given on page 88 as it was probably out of use by then. The large scale 1902 Ordnance Survey Map shows two shafts in this area connected by a 'Pumping Rod' but the position of the 'Pump' is not clear, such a 'rod' would be used to transmit power from a pumping engine to a shaft at a distance.

Wrentnall Mine, Church Pulverbatch, c.1920. This mine was worked as a small mine with about four men from 1890 and again from 1908 to 1916 with up to ten men. Then after takeover by the Wrentnall Barytes Co. this increased to about forty men until the mine closed in 1925. In its later years production averaged about 2,500 tons per year. The photograph does not show any headframes. This was mainly an adit mine with some opencutting of barytes. Processing was carried out in the mill on site.

Cothercott Mine, Pulverbatch *c*.1923. Cothercott Mine was mainly worked by adits but about 1921 a shaft was sunk and minerals raised in kibbles (large buckets) as shown. The mine was opened in 1911 to work a vein of very pure barytes averaging 5.5ft thick but sometimes up to 12ft thick. The mine employed about thirty-five persons and closed in 1928 although it reopened again for a short time in 1934-1935. The barytes was brought from about ten adits and shafts to the mill by the roadside using a small tramway. In 1922 a Geological Survey officer reported that the tramway was 'being extended from the mineworks to connect with the GWR main line near Dorrington, a distance of about four miles'. About one mile of this railway, starting from the mill, was constructed, and a Hunslet saddle tank locomotive, 1ft 10.75in gauge, obtained together with some side-tipping wagons. The photograph shows the headframe and means of emptying the kibbles of ore. (K.C. Lock).

Cothercott Mill, Pulverbatch, 1920s. The barytes was carried from the adits and shafts at Cottercott mine to a mill alongside the road using small tramways. Two of these can be seen arcing around the hillside to the right of the photograph. After treatment the milled barytes was taken away, using the narrow gauge railway to be seen on the opposite side of the road. A photograph of the mill buildings was used by W.R. Cappell in 1930 to provide an illustration purporting to show the mine 'Lostwithin' as described in Mary Webb's book, *The Golden Arrow*. The line of trestles supporting the Huglith to Malehurst barytes ropeway is shown on this photograph crossing the fields on the left. The mill has several barn-like structures with rounded roofs and most appear to have been of timber and sheet construction. (British Geological Survey).

Cothercott Mill 1960s. At the Mill, which opened about 1914, the barytes was washed and handpicked before being ground by milling. In 1922 there were two rotary mills plus four pairs of French Buhr stones , 'while 6 more are being installed'. Perhaps only the first batch were installed because the writer remembers only seeing three pairs, five of the millstones are shown in the photograph.

Cothercott Mine Railway 1940s. The photograph shows joyriders on the remains of the railway, believed to be in the 1940s. The track is said to have been 2ft gauge – slightly wider than the gauge of the secondhand engine – which made it prone to derailment. (Shropshire Records & Research Centre).

Huglith Mine about 1930. This was a highly productive barytes mine which commenced production in 1910, and reached a peak, exceeding 20,000 tons per annum in the 1930s. The mine finally closed in 1945 having produced over 290,000 tons. At most the mine employed sixty five men, in 1935. The photograph shows the mine site, in a woodland setting and with a rope-hauled tramway in the foreground. (K.C. Lock).

Huglith Mine 1930s. A group of miners are shown in their workclothes with the wide brimmed hard hat with a 'gob of clay' and candle that seems still to be standard issue. The buildings on the previous photograph can be identified. (Emily Griffiths Collection).

Rock drilling by compressed air, Huglith Mine 1930s. One of the greatest steps forward in working methods was the introduction of compressed air for drilling shotholes. The first installation was at Tankerville Mine in 1880 then the following year 1881 at Snailbeach. Other mines followed but the miners were not too happy. Rock drilling by machine led to more accidents than by hand and consequently the machine was nicknamed 'the widow maker'. In addition the dust produced caused dust disease or silicosis. The air from these machines, however, aided the mine ventilation considerably.

Huglith Mine about 1930. Tom Hocking, the manager of the mine, stands in the doorway of the mine office and a group of workmen in front. The managers of the barytes mines do not seem to have been accorded the courtesy title of 'captain' normally given to the managers of Shropshire metal mines. Huglith Mine had a succession of owners including T. Butler, Shropshire Mines Ltd, Malehurst Barytes Co. and Laporte Industries. (K.C. Lock).

Huglith Mine c.1930. Loading out material after the blasting of the hanging wall at one of the adit entrances. The vein was not quite vertical as can be seen from the angle of the foot-wall (lower cheek of the vein) on the left. (K.C. Lock).

Huglith Mine c.1930. Reconstructing the headgear arrangements, with Charles Pidwell from Cornwall on the right. He was a mining engineer who came to help reorganise the mine about 1920. The company designed and built all its headgears in timber or steel as required. (K.C. Lock).

Huglith Mine Ropeway 1920s. Up to about 1922 the barytes from the mine was taken, after sorting and washing at the mine, by tramway to a loading place by the nearest road, then by road transport to the railway at Pontesbury. From 1922 the material was sent by aerial ropeway to the railway, then to Malehurst Mill as it became available. At 3.5 miles long this was the second longest ropeway in the area.

This photograph is believed to show the construction of the aerial ropeway loading station. Two trestles can be seen above the heads of the worker standing to the left of the picture. (K.C. Lock),

Left: Huglith Mine Ropeway c.1920. This photograph is believed to show a trestle for the ropeway somewhere near Malehurst Mill, timber trestles were generally used but steel trestles were used near public highways. The ropeway was dismantled in 1949. Malehurst Mill used the standard process for mineral dressing but with the addition of a jaw crusher for crushing large lumps and extra mills for reducing to the size of particle required. (K.C. Lock) Right: Malehurst Mill Ropeway Terminal 1920s. This is believed to be a photograph taken at the terminal of the ropeway within the works area. (Emily Griffiths Collection).

Huglith Mine Shaft Headframe c.1918. Here is shown a wooden headframe with single pulley and, in the foreground, a 'rolley post' or additional support to take the weight and to guide a heavy rope. Huglith Mine was really two mines, No.1 and a No.2. No.1 was adjacent to the old Westcott copper mine and in 1921 had a shaft to 165ft below adit level. No.2 had been opened about 1916 but only got to 80ft depth when the vein disappeared so No.1 was refurbished and kept going until another No.2 was sunk in 1931. Eventually this took over as producing unit. (K.C. Lock).

Sallies Mine Shaft Headframe c.1945. This headframe is made of steel but is of a similar design to the one above. A group of visitors is being taken around the underground workings and several are on the cage. The visitors are covered with a black coloured wet suit, have a modern black helmet and hold helmet-style carbide lamps. (K.C. Lock).

The Sallies Mine 1947. Officials from the mine owners, Laporte Industries Ltd, are inspecting the mine underground. It was opened as a trial in 1937 by the former owners, Malehurst Barytes Co., and produced moderate amounts during the war, afterwards it was developed and employed nearly forty persons. The Mine closed in 1948. The mineral was taken from here to the old Huglith Mine for transfer to the mill by the aerial ropeway. (K.C. Lock).

Miners at Sallies Mine 1946-1947. Towards the end of the mine's life miners gather for a photograph, possibly a celebration as one man is drinking. Back row, left to right: mining student, Harold Betton; Alf Lewis of Pontesbury; Alf Jones; Olwyn Jones; Hebor Rowson; Bill Passant; Alf Lewis; Bill Jones; Tom Clarkson; Arthur Woodman. Centre row: Edmund Betton; Sid Simcox; Eric Davies; Albert Jones (carpenter). Front row: Wilson Morgan; Bill Cowell; Bill Harris; Tommy Garner. (Emily Griffiths Collection).

Sallies Mine 1946. This underground photograph shows overhand stoping for barytes in progress. Compressed-air drills are in use to bore the shotholes and there is a device present for reducing the amount of dust produced. During this time the mines became quite notorious for the incidence of silicosis or 'dust disease' occurring and the miners were encouraged to wear masks. The miners shown at work were; Wilson Morgan, holding drill, George Williams, holding pipe, and in the background Alf Lewis, John Roberts and Jim Harper. (K.C. Lock).

The Sallies Mine 1946. Another view of the overhand stoping method employed at the mine. Illumination was by carbide cap lamps. The mine was very wet with water 'up to your knees'. (K.C. Lock).

Wilderley Copper Mine , near Cothercott Hill Farm, 1917. These photographs are taken from a 'prospectus' promoting this mine and show how far it had been developed. The mine had an adit or level and two shafts 40ft and 70ft deep. Each photograph had its own handwritten caption. This one, 'Entrance – Main Level' shows three of the workers – with cloth caps and drooping moustaches.

Wilderley Copper Mine 1917. This had the caption, 'The morning shift – note method for quick delivery of ore from shaft into shoot connecting with tables below'. The ore was obviously raised in the enclosed headframe, lowered onto a trolley pushed over the shaft to receive it, then disconnected from the rope and pulled out on the trolley. This was quite the normal method for small mines in Shropshire where the trolley was often called the 'tacking plate'. (IGMT).

Wilderley Copper Mine 1917. 'Winch and boiler before covering in'. At this stage the shed was only part covered in, work had not yet started on enclosing the headframe at the back of the photograph, or the engine house. The boiler was vertical and a standard winch was in use as the shaft was very shallow. (IGMT).

Wilderley Copper Mine 1917. 'General view of works, reservoir, engine house and main shaft'. The works indicated above have been completed and two other buildings added, the one immediately below the open door of the pithead building was on the 'site of original adit and dump', it contained 'washing and grading tables'. The long low building was the completed winch house. (IGMT).

Wilderley Copper Mine 1917. 'Washing and grading tables, platform and tramway'. Ore production under way. Various grades have been separated and provision made by tramway for movement to the road. When an inspector was reporting on the mine to consider its status as an essential operation during the Great War, he found a heap of ore at the shaft, from which 6-8 tons of handsorted material had been taken. This it was said represented the full production of the mine before this visit. The mine employed about eighteen persons in 1917-19 but did not continue after this date. (IGMT).

Cornish Engine House believed to be at Westcott Copper Mine 1918. Only one early photograph of a Shropshire copper mine is known. The much reduced ruins of the structures can still be seen. The mine is recorded as being worked between 1866 and 1868 when about 180 tons copper ore was produced, it was again worked 1891-1893 employing about four men. A sketch of the photo by M. Newton is given for clarity.

Road Locomotive 1917. This engine was used to haul lead ore and barytes from the mines of the Shropshire Lead Mines Ltd, particularly The Bog Mine, to the station at Minsterley. The engine was a Foster-Wellington tractor weighing 6 tons. It was of 4 nominal horsepower (approximately 35bhp) and carried extra water tanks. Although one of the smaller tractors in use in the area, this one was capable of hauling 6-8 tons. A newspaper report says that on 8 November 1870 the people of Minsterley were 'on tiptoe with excitement' when a 'Traction Engine and Tender arrived to convey ore for the Pennerley Mining Co.' (K.C. Lock).

Aerial Ropeway, Bog to Minsterley, c.1920. About 1917 Col. Josslyn Ramsden was instrumental in getting a 5.5 mile long ropeway constructed from Bog Mine to near Minsterley Station as an alternative to road transport. Some of the labour was provided by German prisoners of war housed locally in a railway workshop. The buckets on the rope also took the anthracite and coal up to the mine for conversion to gas for the gas engines. The photograph must have been taken somewhere near Tankerville Mine as the square enclosure containing the Buxton Powder Magazine can be seen in the background. The ropeway was dismantled in 1926 or 1927. (Emily Griffiths Collection) Inset: Aerial ropeway junction at Roundhill c.1920. The main Bog to Minsterley ropeway was joined by two feeder ropeways at Roundhill. This is the view at the collecting point but it is not clear how it worked! (Emily Griffiths Collection).

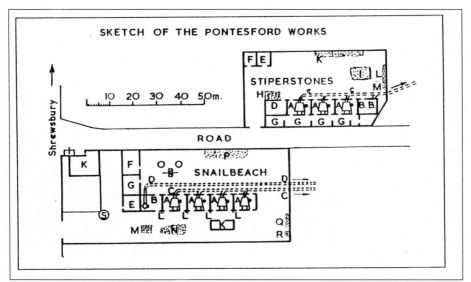

The Pontesford Smelt Houses 1862. The idea behind constructing the smelthouse at least five miles from the lead mines must have been that it was cheaper to transport a small tonnage of lead ore to a large tonnage of coal than vice-versa. The coal was necessary both for the smelting operation and for the associated steam engines. The smelt house owners also acquired local coal mines to provide that coal. However, because of the inferior nature of the coal, this did not work out so well and some better quality coal had to be brought in. The drawings show the layout of both the Stiperstone Co.'s, and the Snailbeach Co.'s smelt houses, with reverberatory furnaces (A) and main flues at (C). These flues led to tall chimneys some distance from the houses. Some idea of the interiors can be seen in the lower drawing.

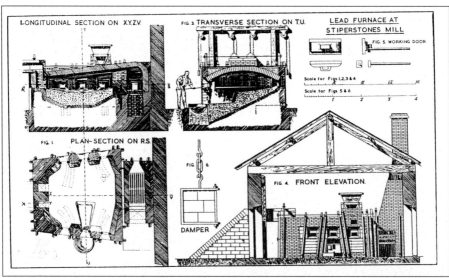

Interior of the Stiperstones Co.'s Smelt House 1862. The smelters were described in detail by M.L. Moissenet, a French Mining Engineer and published in the Annales des Mines in 1862. Recently H.M. Martell and M.C. Gill have translated the paper and clarified the drawings. The two reproduced here are by their kind permission and the full papers can be seen in the Bulletin of the Peak District Mines Historical Society Vol.11, No.6, Winter 1992.

The Stiperstones Smelt House Site 1989. This smelt house lies on the north side of the road to Pontesford Hill and the building shown appears to be based on the original foundations. It contains part of the structure although it has been modified over time and now forms a store.

The Stiperstones Smelt House Site 1989. The inside of the above building showing that, while the roof materials have been changed, the beam structure is similar to that shown opposite. The buildings are of historical importance and the photograph shows visiting historians considering the features.

Minsterley Station, on the Minsterley Branch Line c.1910. This branch of the Shrewsbury &
Welshpool Railway was part of a scheme to take a line to Bishops Castle and Central Wales but
the line was never built beyond this point. The railway and station were opened with great rejoic-
ings in February 1861. This railway had the full support of the Snailbeach Mining Co. and its
management, Mr Job and Mr Harrison, gave speeches at its opening and Mr William Burr of
Kingsland, Shrewsbury, 'Lead Works and Shot Tower Proprietor', presided. Mr Burr's company
was the major user of lead from Snailbeach and expected to be a regular user of the Line. Other
stations on the Branch which received goods from various mines were Pontesbury, Plealey Road
and Hanwood. Passenger traffic on the line ceased in 1951 and goods ceased in 1967.

Pontesford Colliery Pumping Engine House, Nags Head Pit c.1980. This engine room may have
been the last in Pontesford used for its original purpose as it has not been converted into a
dwelling. It is impossible to photograph the full building from the front but the artist, Malcolm
Newton, was able to draw it by 'moving' the tree to the right – an artistic licence.

Seven

The Shrewsbury Coalfields

This coalfield really comprises three separate coalfields each much worked by small pits in the past. The Dryton Coalfield, near Eaton Constantine, is small, isolated and has not been worked for well over a century. The Hanwood Coalfield averages about three miles in width and extends for eighteen miles between Haughmond Hill and the Breidden Hills. The third coalfield at Leebotwood is about thirty square miles in area and is almost square in shape.

The coal seams are all in the Upper Coal Measures which, in this area, rest unconformably upon Pre-Carboniferous rocks. In the two more important coalfields the workable seams, in ascending order, were the Half-Yard Coal (about 1ft 6in thick), the Yard Coal (2ft 9in) and the Thin, Best or Deep Coal (1ft 6in). Associated with these seams there was a bed of limestone about 5ft thick which, prior to 1870, was also mined and used for agriculture.

Coal mining in this area is recorded from at least the early seventeenth century but larger scale development began with the construction of steam engines and the building of lead smelt houses by the lead mining companies in the late eighteenth century. In the mid-nineteenth century these companies still had an interest in some coal mines. In all nine collieries were registered in the coalfield in 1860 and ten in 1884 (after the arrival of the railways). From about this time the mines became few in number but bigger and in 1905 five were registered, employing over 300 persons in total.

In 1920 there were three mines working; Hanwood, Cruck Meole and Moat Hall, all owned by Mr A.N. Fielden, (Arscott & Nobold Collieries owned by J.A. Smallshaw, having just closed) and 373 persons were employed. By 1930 the three mines were combined and numbers employed had fallen to 200 – they then continued to fall slowly until working stopped in 1940. One small mine at Castle Place, Pontesbury, was opened after the Second World War but this had closed by 1947.

Mining conditions in the Coalfields were dangerous, being both wet and subject to bad air. Between 1850 and 1900 twenty-one fatalities were recorded; perhaps the worst were an inrush of water at Pontesbury mine in 1856, which killed three, and the breakage of a winding rope at Shorthill Mine in 1882, that resulted in the death of two. Falls of roof, tunnel sides or the coal face and falling down shafts were the most usual causes of death.

There are three surviving engine houses in Pontesford; they all seem to be of the same period but little is known of them at present. None seem to have been used for their original purpose since the mid nineteenth century; two have been converted into dwellings and one survives in ruins. This latter building is shown on the opposite page.

Pontesford Colliery Pumping Engine House. The collieries of this area were used to supply the fuel for the engines and smelters associated with the adjacent metal-mining area. As early as 1775 a 'fire engine' is recorded in the vicinity and there is good documentary evidence of several other engines used at Pontesford although to which mine the information relates has not yet been determined. It is known, however, that this house has been a dwelling since before 1840 as a former occupier's father was born in the house in 1843. The building is divided into three storeys and the engine pit is used as a cellar. The pumping shaft is about 6ft from the lever wall but is now filled and capped.

Pontesford Colliery Pumping Engine House, 1980s. The site is shown on an old plan as 'fire engine'. The term 'fire' was used to distinguish it from engines with other forms of energy. The house is now part of a row of cottages. As indicated by the presence of three such engine-houses, water was always a problem of these mines. Three miners were drowned in 1856 when water burst in on them in one Pontesbury mine.

Moat Hall Colliery, late 1890s. This colliery was opened prior to 1850; in 1860 it was owned by 'T. Jones' and in 1875, another owner, Mr Shorthouse, sank a new shaft about 150 yards deep. The mine continued to work under the Shorthouse family until 1919 when it was acquired by Mr A.N. Fielden who later closed it but continued to work the seams from his Hanwood Colliery. The photograph shows the mine workforce standing in front of the wooden headframe. (K.C.Lock).

Moat Hall Colliery 1920s. This does not seem to be the same shaft headframe as shown above but Moat Hall had several shafts. (In 1891 they were working the Half Yard Seam at No.6 and No.7 shafts each about 110 yards deep). These shafts were connected to the road near Annscroft by a tramway. During this century the mine employed on average thirty men underground and fifteen on the surface. The surface arrangements are shown to be simple, a steam winder, a rolley post to support the rope and a traditional 'pillbox' for the banksman to shelter in. The miners on this photograph are believed to be Ernest Lewis, Ern Challinor and Jack Tipton.

Cruckmeole Colliery c.1910. The colliery was probably opened mid-nineteenth century, by 1891 the shafts were 120 yards deep and it was owned by S. Atherton. In 1905 sixty men were employed and this increased to about seventy during the First World War. In 1919 the mine was bought by Mr Fielden and coal was brought here by tramway from his Hanwood Colliery for screening. The photograph probably dates from this time. Coal production ceased at Cruckmeole in 1926 but the shaft continued to be used for ventilation and pumping until nearby Hanwood Colliery closed in 1940. Pumping was by beam engine until 1919, when Evans & Morgan steam pumps were installed and finally electric pumps were used. The mine had a brickworks adjacent to it which used some clay spoil from the heaps. A new continuous kiln was put in in 1921-1922 and the works closed in 1941. (K.C. Lock).

Cruckmeole Colliery, possibly earlier than the above. The cage has arrived at the surface and contains a wagon of coal. The surface arrangements are similar to the photograph of Moat Hall overleaf but there are differences in the rolley post structure and the ladderway is on different legs of the main headframe. (K.C. Lock).

Hanwood Colliery c.1920. This mine was sunk in the 1870s and had only one workable seam', the Half-Yard, which was less than 2ft thick. The haulage roads had to be 'ripped' to give extra 'headroom', the waste being put into the goaf. On the face, sledges were used, pulled by a chain between the haulage-boys' legs, and this continued until the late 1930s. By 1891 the shaft was 420ft deep and ventilation was by 'hot steam' being exhausted into the up-shaft air thus increasing the current. In 1920 the mine was taken over by Mr Fielden along with Cruckmeole Colliery and the complex was operated as the Hanwood and Moat Hall Collieries Ltd. Between 1921 and its closure in 1940 the numbers employed gradually fell from over 300 to less than 100 but for much of the time production averaged about 25,000 tons of coal per year. The photograph gives a general view of the mine before a second pulley was added, for use with a counterweight, in the mid 1920s. (K.C. Lock).

Hanwood Colliery 1920s. In 1921 a power station was built at Cruckmeole Shaft (which was now part of Hanwood) to supply the mine with electricity. Steam was used to drive two Williams – Robinson generator sets rated at 200kW each. These are shown in the photograph. The power was used for driving an electric haulage underground, for pumping and for lighting but not apparently for winding in the shafts as steam winders were used until closure. (K.C. Lock).

Hanwood Colliery haulage engine 1920s. The 75hp electric haulage engine used underground is shown but there was still work for three ponies in the side roads. The photograph shows the engine with Charles Cooper and Bill Hughes.

Hanwood Colliery c.1930. This photograph shows 'Charlie' the pony being prepared for lowering down the shaft. Also on the photograph are, left to right: W. Challinor; Edwin Thomas (horse-keeper); George Cooper; Charles Cooper; and Mr Bolton (manager). (K.C. Lock).

Hanwood Colliery c.1930. Two miners are moving a sledge or 'dan' of coal along the coal face in the 1930s. Normally boys would have been used to haul the sledges with chains but, with the advent of motor cycles giving easier access to alternative work, the boys left the mines and the older miners had to move the sledges themselves. The miners were Percy Mansell and Walter Challinor. Note the candle stuck in clay for illumination. (K.C. Lock).

Hanwood Colliery c.1930. Setting a support in a tunnel leading to a coalface, the roof of the coal seam is having to be removed to make 'headroom'. The dates chalked on the coal indicated when the work was last measured or inspected, in this case 11, 13 and 14 April, but no year is given. The other mark is the initials of the deputy or manager. It is believed that the miners are the same as named above. (K.C. Lock).

Hanwood Colliery c.1930. A group of miners standing in front of the mine cage with the tramway and a wagon in front of them. The miners were part of a close-knit community. During the 1920s there were some disputes but the managing director, Mr A.N. Fielden, offered a profit sharing scheme and various other 'benefits' for his workers. Among the workers were J. Newnes, C. Cooper and W. Potts.

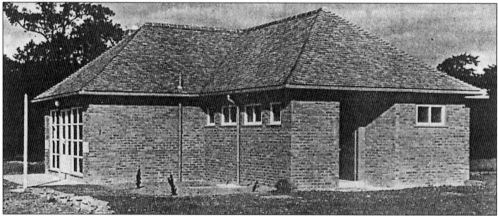

Hanwood Miners' Welfare Pavilion 1938. This pavilion was built with money from the national Welfare Scheme where mine owners and coal owners had to put 1d or 2d per ton of coal produced to be used to benefit the miners. Hanwood miners applied for and got their pavilion which cost £1000 and had a central room big enough for social functions and educational classes. A report says 'the paint was scarcely dry before the first lecture was delivered, the torch of learning must needs burn brightly for the electricity supply is not yet available!'

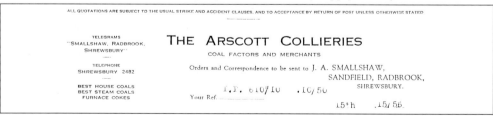

Arscott Collieries headed paper, 1956. Arscott Collieries actually closed in the 1920s but the company has continued to operate as 'Coal Factors' or distributors since.

Eight

The Coal and Other Mines of North West Shropshire

Although primarily an agricultural area there have been a variety of important mining industries in this part of the county. One mine, Ifton, eclipsed all mines not only in this area but in Shropshire in the numbers employed but all the others were small both in production and surface employment. The area's surface mining industry, however, continues mainly in the form of quarrying for aggregates but some peat and monumental stone is also produced at the present time.

The North Shropshire, or Oswestry Coalfield, is really an extension of the Denbighshire Coalfield and, in Shropshire, covers an area of about sixteen square miles. Although this is only a small part of the coalfield, proportionally it has been quite productive. The last large mine, at Ifton, closed as recently as 1968. The coalfield contains seven important seams but the Two-Yard, Quaker, Powell and Main Coal seams have been the most important in recent years.

Copper has been found, as green malachite in the Triassic Sandstones of North West Shropshire but, except for the occasional shaft or adit, has left few surface reminders. Visitors have recorded seeing mines here since the seventeenth century but it was in the nineteenth century that they really made their mark. The mines of Eardiston and Clive are well known but others can be found at Pimhill, Wixhill and Hawkstone. They have occasionally produced other minerals too including cobalt, vanadium and lead.

Significant quantities of copper and also lead are believed to have been obtained from the limestone at Llanymynech. Some of these workings could be Roman or earlier. Other areas where mining has taken place include Llynclys Hill, Crickheath and perhaps the heavily pitted area around Moelydd, of which little is known. The surface workings for peat and Grinshill Sandstone, while not underground, have also a very long and important history; they cannot be ignored.

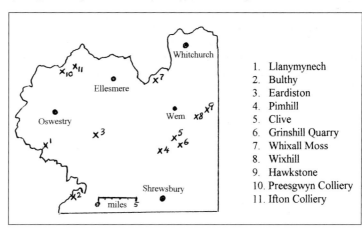

1. Llanymynech
2. Bulthy
3. Eardiston
4. Pimhill
5. Clive
6. Grinshill Quarry
7. Whixall Moss
8. Wixhill
9. Hawkstone
10. Preesgwyn Colliery
11. Ifton Colliery

Location of Mines of North West Shropshire.

Preesgweene Colliery 1873. This colliery, near Weston Rhyn, had a number of shafts and eventually reached a depth of about 500ft. It was, as can be seen from the number of employees in the photograph, quite a substantial undertaking, but eventually closed in 1891. One surprising feature is the large number of obviously very young children employed, shown seated on the ground at the front of the picture. The principal coal seams worked were the Top and Lower Yard, Two Feet, Five Feet and Six Feet Seams. Donkeys commonly served on colliery surfaces but occasionally also underground, in 1900 there were twenty-six donkeys in use at Shropshire mines. (Shropshire Records and Research Centre).

Ifton Mine, St Martins 1966. The mine was originally two separate concerns, Brynkinalt Colliery and Ifton Rhyn Colliery. Brynkinalt shafts were 165 yards deep and were sunk in the second half of the last century. The Ifton Rhyn shaft, 1.5 miles east of Brynkinalt, is of a similar date but prior to deepening was only 135 yards deep. A tunnel driven from Brynkinalt connected with Ifton shaft in 1923 and in 1928 Brynkinalt shafts ceased to be used for coalwinding. At this time the mine employed about 1,357 men. By far the biggest mine ever to operate in Shropshire. The mine was Nationalised in 1947 and shortly afterwards the nearby Black Park Mine was closed and its workings were incorporated into the former. Six workable coal seams were available spread vertically over 300ft of strata. On Saturday, 23 November 1968, the mine ceased production and 1,234 jobs were lost.

Ifton Colliery Underground 1960s. Ifton Colliery was a highly mechanised mine in the 1960s as this picture shows. A drum-shearer coal cutter, riding on a heavy duty 'panzer' conveyor structure, is being hauled along the face using a chain. The man is working on the 'waste' side of the conveyor and the roof above is supported by hydraulic props and steel bars. The conveyor was pushed forwards behind the cutter by hydraulic rams situated on the floor. (Shropshire Mines Preservation Trust).

Ifton Colliery Underground 1963. Empty wagons are being hauled up a bank by a chain so that they can gravitate to a point beneath the conveyor head to be loaded. After loading they were connected together and hauled out of the mine by rope or, later, locomotive.

Ifton Colliery 1968. The colliery was closed by the NCB in 1968. Its closure was the result of geological problems, a fall in 'markets' for coal and other difficulties. At the end of the last working shift a miners' safety lamp was presented to the leader of the miners' union, Sid Davies.

Ifton Colliery Miners' Institute 1932. This Institute was built at a cost of £7,500, with furnishings £950 extra. The money came from the Miners' Welfare Fund built up by a levy on the mine owners and coal owners of so much per ton produced. An outdoor recreation area with bowling green, tennis courts and putting course had already been provided.

The main hall seated 444 persons and had several anterooms, there were also rooms for billiards, games, reading, library and a kitchen. (The games and reading rooms could be put together to form a refreshments area). The building was of brick with slate roof and a generator was provided to give electricity for heating, cooking and lighting, there being 'no electric mains within several miles'.

Llanymynech Mine Entrance 1986. This mine, in limestone, has produced copper and lead ores. It is situated about five miles south-west of Oswestry and the site is now mainly a golf course. Thirty Roman coins were found, in 1964, buried in the mine, and old tools and other artifacts have also been brought out. The mine is likely, therefore, to be Roman or earlier. It was also recorded in the twelfth century and at other times and during the mid-nineteenth century. Shropshire Caving & Mining Club members are shown standing in the Entrance Chamber prior to an underground trip. (Kelvin Lake/IARecordings).

Llanymynech Mine Underground 1986. The copper miners' workings in the limestone deposits are very irregular as can be seen in this photograph. The various working chambers are entered through irregular shaped holes in the roof, side, or, as in this case, the floor. Small pillars of rock have been left either for support or because they appeared to contain no ore and were not worth taking. Two members of the Shropshire Caving & Mining Club, E. Thorpe and M. Moore, are entering on either side of the pillar shown. (Kelvin Lake/IARecordings).

Clive Copper Mines 1986. There are a number of shafts in and around the village of Clive and exploration of the workings has revealed what must be the largest copper working in Shropshire. There is documentary evidence of mines here from the early eighteenth century to mid-nineteenth century. The main period of working seems to have been from 1862 to 1870 when, at times, up to thirty men were employed but no record of production has been found.

In recent years the shaft has been used for pumping water, formerly a steam driven pump was used but more recently pumping was carried out by an electric powered three-throw ram pump. The pictures show the view down the shaft, three pump rods can be seen with a ladderway and the 'rising main'. (Kelvin Lake/IARecordings).

Hawkstone Copper Mines 1974. Mines for copper are recorded in this vicinity from the seventeenth to the mid-nineteenth century but there never seems to have been much success. The last working was probably at Wixhill in the 1860s. Near some of the mine shafts are 'follies' and 'grottoes' cut out of the sandstone rock which could well be the work of some of the miners. The ones shown are in a Country Park, easily accessible and well worth a visit.

Eardiston Mine Entrance 1984. This was a mine for copper ore found in the Triassic Sandstone. As well as having one major adit the mine had at least one shaft about 100ft deep with a steam engine for winding or pumping. Some ruins of the engine house still remain on site. The mine was worked on several occasions in the nineteenth century but especially in the 1840s and 1860s. The picture shows the 'modern' shape of the entrance to the adit with Edwin Thorpe providing the 'scale'. It is unlikely that the mine produced much copper ore although some samples taken were rich in the mineral. Some drilling for copper ore was being carried out near here in 1966. (Kelvin Lake/IARecordings).

Pim Hill Copper Mine – main tunnel, 1990. The old copper miners' tunnels are very different in the sandstones to those in the limestone – here they are spacious and regular. Two explorers are studying the marks made by the picks of the miners as they hand-cut the tunnels. There are at least three shafts at this mine. In the early 1700s the mine was worked by Abraham Darby (of Coalbrookdale Co. fame) but the main period of work was the 1860s. The mines were also searched for cobalt ore in the early twentieth century. (Kelvin Lake/IARecordings).

Grinshill Sandstone Quarries 1898. The Grinshill quarries have been worked for building stone for at least ten centuries. They produced the stone for many important buildings in the county and beyond. Among these are Haughmond Abbey, the churches of St Mary and St Julian in Shrewsbury, Attingham Hall, Hawkstone Hall and more recent buildings. In 1858 the stone was advertised at '6d per cubic foot', in 1896, the quarries employed thirty-three persons and in 1911, nineteen persons. The quarries are still in operation. In this photograph one of the traditional steam powered derrick cranes can be seen in operation. ('The Quarry' journal 1898).

Grinshill Sandstone Quarries 1898. Workmen are shown cutting out the blocks of stone in the main quarry 140ft deep. The cuts or channels were 9-12in wide and 3-4ft deep and were made by hand using picks up to 3ft long. Pinchbars were then used to detach the blocks which were then lifted by handcranes into a position to be raised to surface by one of two steam derrick-cranes. The largest stones sold were 12ft by 5ft by 5ft. (The Quarry, 1898). Inset: A Grinshill Pick.

Peat Working on Whixall Moss. Whixall Moss, extending into Wales, where it is known as Fenns Moss, is a lowland raised bog with an extraction history going back many centuries although it has no documented history before the sixteenth century. There have been traditional or 'turbary' rights to work the moss with their own terminology and control structures since this time but most were lost during the enclosures of 1710 and 1823. Local peatcutters continued to work the Moss on a rented 'acre' by 'acre' basis up to the 1960s when larger companies took over. The photograph shows an area of diggings on the Moss carried out by traditional tools and in the local style.

Railways on the Moss 1980. Rail transport has been used on the Moss since before the 1850s but most of it has consisted of hand-pushed wagons. In 1919 a petrol-loco was introduced, and at times since up to three other locos have been used. The loco shown is believed to have been one of the earliest and it, and some wagons, have now been preserved on the site. Loco haulage was used on parts of the Moss until 1970. In 1990, a great effort by all the authorities concerned with the environment enabled English Nature and its Welsh equivalent to take control of the Moss and under them only a handful of traditional 'peatmen' are allowed to operate. (Kelvin Lake/IARecordings).

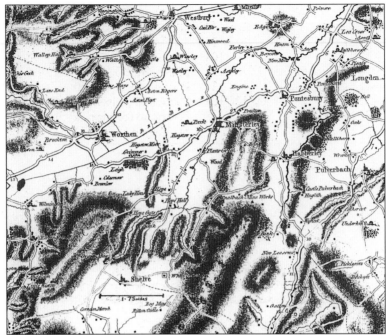

Left: Part of Robert Baugh's Map of Shropshire 1808. This part of Baugh's map covering the Shelve-Minsterley area gives a good indication of mining activity even at this early date. 'Snalbach' Mine works', 'Bog Mine' and 'Gravels Mine' are all named but Pennerley Mine is shown as 'White Grit Mine'. The early Boulton & Watt steam engine at Malehurst between Minsterley and Pontesbury is shown as 'Engine' with two coal pits (indicated by dark round spots) nearby. (IGMT).

Index of Mine locations

Note – Page numbers and Ordnance Survey Grid References are given. (some are general locations only).

Bibliography

Aikin A. Journal of a Tour through North Wales and Part of Shropshire; with observations in Mineralogy and other branches of Natural History, London. 1797.

Allbutt M. & Brook F. 'The Snailbeach Lead Mine', Ind.Arch.Soc. Journal, Portsmouth College 1969.

Anon. 'The Bridge and Cureton Company Quarries, Grinshill', The Quarry. Sept 1898.

Berry A.Q.et al. Fenns and Whixall Mosses, Clwyd County Council. 1996.

Brook F. & Allbutt M. The Shropshire Lead Mines, Moorland Publishing. 1973.

Brown I.J. The Mines of Shropshire, Moorland Publishing. 1976.

Brown I.J. ' Preserving mining's past - the Bog Mine', Mining Environmental Management.1994.

Brown I.J. 'New life for Snailbeach Mines', Mining Magazine. March 1995.

Burt R. et al The Mines of Shropshire and Montgomeryshire. Exeter. 1990.

Davies T.J. The Engine Houses of the South Shropshire Mines, privately printed 1969.

Francis P. Price J. and Yapp J. Never on Sunday - Memories of the Stiperstones Mining Communities, Shropshire Mines Trust. 2000.

Gayden A.T. & Lawson J.B. 'A history of Pontesbury', from Victoria County History Vol V111 1962 reprint 1982.

Geological Survey. Special Rep. Vol. XX11 Lead and Zinc Ores. 1922.

Geological Survey. Special Rep. Vol. 11 Barytes and Witherite 1915, 1916, revised 1922.

Geological Survey. Special Rep. Vol. XXX Copper Ores. 1925.

Geological Survey. Special Rep. Bulletin No. 14, The West Shropshire Mining Region. 1958.

Geological Survey Wartime Pamphlet No. 46, Barium Minerals, published 1945.

Minutes of Evidence to the Commission on the Condition of Mines (Kinaird Commission) Vol. 7. 1863.

Moissenet M.L. 'Lead Mining & Smelting in the Snailbeach District' (in French), in Annales des Mines, Paris, 1862. (Abstract in English), 'Mining and Smelting Magazine' 1862).

Morton G.H. 'The Mineral Veins of Shelve' Proc. Liverpool Geol.Soc., 1, and separately printed, Liverpool. 1869.

Murchison R.I. The Silurian System, 2 vols. London. 1839.

Thomas R.D. Industries of the Morda Valley. 1939 (reprinted Shropshire Libraries 1978).

Thompson D.B. A Guide to the history and geology of quarrying in Corbet Wood, Grinshill. 1995.

Tonks E.S. The Snailbeach District Railway. 1950, revised 1974.

Substantive articles have also been produced in various journals, including the following;
In Caradoc and Severn Valley Field Club Transactions.

> Turner E.A. Lead Mining in Shropshire 194? reprinted by SCMC 1968.
>
> Yelland W. Mineral Veins of Shropshire. 1900.
>
> Yelland W. The Working of a Lead Mine. 1911 reprinted by SCMC 1968.

In Shropshire Magazine.

> Newton M. Snailbeach - Shropshire's greatest lead mine Oct. 1978.
>
> Newton M. The Mines at Snailbeach. Feb. 1972.
>
> Newton M. A look at what's left of the Shropshire lead mines. Nov. 1974.
>
> Lock K.C. When Shrewsbury had its own coalfield. Jan. 1970.
>
> Brown I.J. Pontesford Mine Engine House. April 1976.
>
> Clayton Jones H . Shropshire's Mining Outpost - the three pits of St Martins March 1960.

In publications of the Peak District Mines Historical Society, Bulletin, Mining History

> Brown I.J. Saving the Snailbeach Mining Area from Land Reclamation 1979.
>
> Brown I.J. Burgam, Shropshire's last working metal mine 1957 - 63. 1990.
>
> Brown I.J. The buildings and equipment used at Pennerley Mine. 1993.
>
> Brown I.J. Snailbeach Mine and the disaster of 1895. 1995.
>
> Carlon C.J. Eardiston Copper Mine. 1981.
>
> Martel H.M. & Gill M.C. Lead Smelting in Welsh Furnaces at Pontesford. 1992.
>
> Sargeant W.A.S. Pyromorphite from the mines of West Shropshire. 1967.

In publications of the Northern Mines Research Society , British Mining

> Allbutt M. & Brook F. The Snailbeach Mining Co. 1767 - 1911.
>
> Carlon C.J. Gallantry Bank and other copper mines in the Shropshire and Cheshire Basin. 1981.
>
> Chapman N. The White Grit Mine. 1984.
>
> Davis R.V. A brief account of the Geology, History and Mechanisation of the Snailbeach Mine. 1968.

In Special Accounts of the Shropshire Caving and Mining Club

> Adams D.R. Survey of the Metal Mines of S.W.Shropshire (No.2) 1962 revised in 1968, 1979 and 1992.New edition edited by Pearce A.J. 1996.
>
> Adams D.R. Survey of the Llanymynech Ogof (No.8) 1970. New edition edited by Pearce A.J. 1992.
>
> Brown I.J. Snailbeach Lead Mines - the Surface Remains (No. 17) 1993.
>
> Davies T.J, Newton M .and Pearce A.J. Mining Remains in SW Shropshire (No.18) 1993.

Other publications of the Shropshire Caving and Mining Club, since 1960; Yearbook, Journal, Newsletter and Quarterly Journal 'Below' contain numerous articles on the mines of West Shropshire.

Details of the availability of the above publications can be obtained at local libraries or from the National Association of Mining History Organisations (see page 2).